Indian
Ocean

YOKOHAMA

OWN

N ˙OCEAN

ROSS
SEA

MY FAVOURITE TALES OF THE SEA

My
Favourite Tales
Of The Sea

chosen by
SIR ALEC ROSE

NAUTICAL PUBLISHING COMPANY
ADLARD COLES   ERROLL BRUCE   RICHARD CREAGH-OSBORNE
Captain's Row · Lymington · Hants

in association with
GEORGE G. HARRAP & COMPANY LTD.
London · Toronto · Wellington · Sydney

© 1969 by Sir Alec Rose

Standard edition SBN 245 59898 7

First published in Great Britain 1969 by
NAUTICAL PUBLISHING COMPANY
Captain's Row, Lymington, Hampshire

Composed in 11 on 13 pt Monotype Baskerville
and made and printed in Great Britain by
Cox & Wyman Ltd., London, Fakenham and Reading

ACKNOWLEDGEMENTS

My thanks are due to Mr Peter Scott for his personal permission to use part of his father's journals, which include some extracts from Captain Scott's letters home. In the same way I should like to thank Mrs Helen Tew for her personal permission to republish part of *Crossing an Ocean*, by her father, Commander R. D. Graham; and also its publishers, William Blackwood and Sons Ltd., for their permission. Equally I am grateful for the personal permission of Brigadier Miles Smeeton to use a tale taken from his book, *Once is Enough*, and also to its publisher, Rupert Hart-Davis. Rupert Hart-Davis has also kindly given permission for extracts to be taken from *Across Three Oceans*, and I am grateful to the author, Conor O'Brien.

My thanks also go to J. M. Dent and Sons, publishers of *Youth* by Joseph Conrad, besides the Trustees of the Joseph Conrad Estate. The yarns by Hilaire Belloc in *The Cruise of the Nona* and in *Conversations with a Cat and Others* are reprinted by permission of A. D. Peter and Company.

The Bodley Head Ltd. has kindly given permission for the inclusion of a passage from *The Venturesome Voyages of Captain Voss* by John Claud Voss. William Heinemann Ltd. have generously stated that they are happy to give permission for the use of the story from *South* by Sir Ernest Shackleton. Jonathan Cape Ltd. have kindly given permission for reprinting the piece from *The Cruise of the Teddy* by Earling Tambs. Also I would like to thank *The Times* for permission to reproduce *Man Overboard*.

Then I should like to thank Christopher Graham for helping me to narrow down my choice to the tales included in this book; also Adlard Coles and Erroll Bruce, my publishers, for their assistance in tracing copyright ownership and giving advice generally.

Finally, I am deeply grateful to those who taught me to

enjoy reading about the sea and about adventure; most of all
to my sister Muriel, who used to read exciting tales to me when
I was too delicate to join in the normal fun and games of small
children, and could only dream that perhaps one day I might
venture forth over the seas myself.

CONTENTS

Introduction

When I was young, I read many adventure stories. I do not suppose that I was different from other boys in this respect; but, because I went through a period when I was not allowed to play games, I had more time to read than many of my contemporaries. Of all the stories that I read it was always sea adventures that most fascinated me. I seemed to feel closer to sailors than to any of the other heroes. I could imagine myself in their situations and dealing with all their difficulties more easily than if I were crossing deserts, struggling up mountains or cutting my way through jungles.

As the years passed, I became as strong as any other young man, but I continued to read about the sea. It was then that I began to read as much for information as for the pleasure of the actual story. I wanted to learn as much as possible about the sea. Every amateur yachtsman owes a great deal to the professional and amateur sailors who have gone before him and who have recorded their experiences as they voyaged about the world. Their accounts encourage him and feed his imagination; but, more usefully, they describe the dangers and hazards that may be met with on a long voyage in strange waters.

During my trips across the Atlantic and round the world, I was usually uncomfortable, sometimes in danger and often frightened, but I also had moments of intense pleasure and happiness. These I have written about elsewhere. In this book I am concerned with the personal experiences of others. I have chosen extracts from stories that I read long ago, but I have added others from more recent books. I have gone back into the past, to the early days of sail, because it is only when I look back that I realize what a wonderful thing the modern sailing ship has become. I marvel at the unbelievable bravery and endurance of people like Anson and Shackleton, at the calm and competent seamanship of Slocum, and at the determination and improvisation shown by the Smeetons and John Guzzwell in saving their ship to get themselves and their cat into safety after being capsized and dismasted.

I hope you enjoy these short extracts from this cross-section of sea stories, and that they will serve to make you ponder on the might of the sea and the amazing ability of man to endure and surmount fearsome difficulties.

Alec H. Rose.

I
Brought Safe to Land

The voyage of the *Mayflower* is an historical event, known to
everyone in Britain; but few have read a description of the voyage
and how, when they were brought safe to land, those aboard the
Mayflower fell upon their knees and blessed the God of Heaven who
had brought them over the vast and furious ocean.

A. R.

from 'HISTORY OF PLYMOUTH 1590–1657'
by WILLIAM BRADFORD

ALL things being now ready, and every business dispatched,
the company was caled togeather. Then they ordered and
distributed their company for either shipe, as they conceived
for the best. And chose a Governour and 2 or 3 assistants for
each shipe, to order the people by the way, and to see to the
dispossing of there provissions, and shuch like affairs. All which
was not only with the liking of the maisters of the ships, but
according to their desires. Which being done, they sett sayle
from thence aboute the 5th of August.

Being thus put to sea they had not gone farr, but Mr Reinolds
the master of the leser ship complained that he found his ship
so leak as he durst not put further to sea till she was mended. So
the master of the biger ship (caled Mr Joans) being consulted
with, they both resolved to put into Dartmouth and have her
ther searched and mended, which accordingly was done, to
their great charg and losse of time and a faire wind. She was
hear thorowly searcht from steme to sterne, some leaks were
found and mended, and now it was conceived by the workmen
and all, that she was sufficiente, and they might proceede

without either fear or danger. So with good hopes from hence, they put to sea againe, conceiving they should goe comfortably on, not looking for any more lets of this kind; but it fell out otherwise, for after they were gone to sea againe above 100 leagues without the Lands End, houlding company togeather all this while, the master of the small ship complained his ship was so leake as he must beare up or sinke at sea, for they could scarce free her with much pumping. So they came to consultation againe, and resolved both ships to bear up backe againe and put into Plimmoth, which accordingly was done. But no spetiall leake could be founde, but it was judged to be the generall weaknes of the shipe, and that shee would not prove sufficiente for the voiage. Upon which it was resolved to dismise her and parte of the companie, and proceede with the other shipe. The which (though it was greeveous, and caused great discouragemente) was put in execution. So after they had tooke out shuch provission as the other ship could well stow, and concluded both what number and what persons to send bak, they made another sad parting, the one ship going backe to London, and the other was to proceede on her viage. Those that went bak were for the most parte shuch as were willing so to doe, either out of some discontente, or feare they conceived of the ill success of the vioage, seeing so many croses befale, and the year time so farr spente; but others, in regarde of their owne weaknes, and charge of many yonge children, were thought least usefull, and most unfite to bear the brunte of this hard adventure; unto which worke of God, and judgmente of their brethern, they were contented to submite. And thus, like Gedions armie, this small number was devided, as if the Lord by this worke of his providence thought these few to many for the great worke he had to doe. But here by the way let me show, how afterward it was found that the leaknes of this ship was partly by being overmasted, and too much pressed with sayles; for after she was sould and put into her old trime, she made many viages and performed her service very sufficiently, to the great profite of her owners. But more espetially, by the cuning and deceite of the master and his company, who were hired to stay a whole year in the cuntrie, and now fancying dislike and fearing wante of victeles, they ploted this strategem to free them selves; as afterwards was knowne, and by some of them confessed. For they apprehended that the greater ship, being of force, and in whom most of the provissions were stowed, she would retayne enough for her selfe, what soever became of them or the passengers; and indeed such speeches had been cast out by some of them; and yet, besides other incouragements, the cheefe of them that came from Leyden wente in this shipe to give the mast contente. But so strong was self love and his fears, as he forgott all duty and former kindnesses, and delt thus falsly with them, though he pretended otherwise.

These troubles being blowne over, and now all being compacte togeather in one shipe, they put to sea againe with a prosperus winde, which continued diverce days togeather, which was some incouragemente unto them; yet according to the usuall maner many were afflicted with seasicknes. And I may not omite hear a spetiall worke of Gods providence. Ther was a proud and very profane yonge man, one of the sea-men, of a lustie, able body, which made him the more hauty; he would allway be contemning the poore people in their sicknes, and cursing them daily with greevous execrations, and did not let to tell them, that he hoped to help to cast halfe of them over board before they came to their jurney's end, and to make mery with what they had; and if he were by any gently reproved, he would curse and swear most bitterly. But it pleased God before they came halfe seas over, to smite this yonge man with a greeveous disease, of which he dyed in a desperate maner, and so was him selfe the first that was throwne overbord. Thus his curses light on his owne head; and it was an astonishmente to all his fellows, for they noted it to be the just hand of God upon him.

After they had injoyed faire winds and weather for a season, they were incountred many times with crosse winds, and mette with many feirce stormes, with which the shipe was shroudly shaken, and her upper works made very leakie; and one of the maine beames in the midd ships was bowed and craked, which put them in some fear that the shipe could not be able to performe the voiage. So some of the cheefe of the company, perceiveing the mariners to feare the suffisiencie of the shipe, as appeared by their mutterings, they entred into serious consulltation with the master and other officers of the ship, to consider in time of the danger; and rather to returne then to cast them selves into a desperate and inevitable perill. And truly ther was great distraction and differance of oppinion amongst the mariners them selves; faine would they doe what could be done for their wages sake, (being now halfe the seas over) and on the other hand they were loath to hazard their lives too desperately. But in examening of all oppinions, the master and others affirmed they knew the ship to be stronge and firme underwater; and for the buckling of the maine beame, ther was a great iron scrue the passengers brought out of Holland, which would raise the beame into his place; the which being done, the carpenter and master affirmed that with a post put under it, set firme in the lower deck, and otherways bounde, he would make it sufficiente. And as for the decks and uper workes they would calke them as well as they could, and though with the workeing of the ship they would not longe keepe stanch, yet ther would otherwise be no great danger, if they did not overpress her with sails.

So they commited them selves to the will of God, and resolved to proceede. In sundrie of these stormes the winds were so feirce, and the seas so high, as they could not beare a knote of saile, but were forced to hull, for diverce days togither. And in one of them, as they thus lay at hull, in a mighty storme, a lustie yonge man (called John Howland) coming upon some occasion above the grattings, was, with a seele of the shipe throwne into the sea; but it pleased God that he caught hould of the top-saile halliards, which hunge over board, and rane out at length; yet he held his hould (though he was sundrie fadomes under water) till he was hald up by the same rope to the brime of the water, and then with a boathooke and other means got into the shipe againe, and his life saved; and though he was something ill with it, yet he lived many years after, and became a profitable member both in church and commone wealthe. In all this viage ther died but one of the passengers, which was William Butten, a youth, servant to Samuell Fuller, when they drew near the coast. But to omite other things, (that I may be breefe,) after longe beating at sea they fell with that land which is called Cape Cod; the which being made and certainly knowne to be it, they were not a little joyfull. After some deliberation had amongst them selves, and with the master of the ship, they tacked aboute and resolved to stande for the southward (the wind and weather being faire) to finde some place aboute Hudsons river for their habitation.

But after they had sailed that course aboute halfe the day, they fell amongst deangerous shoulds and roring breakers, and they were so farr intangled ther with as they conceived them selves in great danger; and the wind shrinking upon them withall, they resolved to bear up againe for the Cape, and thought them selves hapy to gett out of those dangers before night overtooke them, as by Gods good providence they did. And the next day they gott into the Cape-harbor wher they ridd in saftie. A word or too by the way of this cape; it was thus first named by Capten Gosnole and his company, Anno: 1602, and after by Capten Smith was caled Cape James; but it retains the former name amongst sea-men. Also that pointe which first shewed those dangerous shoulds unto them, they called Pointe Care, and Tuckers Terrour; but the French and Dutch to this day call it Malabarr, by reason of those perilous shoulds, and the losses they have suffered their.

Being thus arived in a good harbor and brought safe to land, they fell upon their knees and blessed the God of heaven, who had brought them over the vast and furious ocean, and delivered them from all the periles and miseries therof, againe to set their feete on the firme and stable earth, their proper elemente.

2

Crossing an Ocean

Emanuel was a cutter only 30 ft in overall length, and when she set
off from Ireland in 1933, few smaller craft had crossed the Atlantic
Ocean; indeed only three other small craft were then known to have
sailed from Europe to America by the northern route since the days
of the Vikings, whose open boats were a good deal longer.

This extract describes the last few days of his voyage, including
the landfall, which is always a significant experience after a long
voyage.

On some of his other long voyages in little *Emanuel*, Commander
Graham was accompanied by his daughter, Mrs Tew, to whom I
am grateful for permission to use this extract.

A. R.

from '*ROUGH PASSAGE*'
by COMMANDER R. D. GRAHAM

JUNE 12th found *Emanuel* running to the westward before a
strong wind and heavy sea. I had now reached the area in
which ice might be encountered, and it was advisable to have
some after-canvas ready to set at a moment's notice. There was
too much wind for the whole mainsail, which should have been
partially rolled up before lowering, and I considered bending
the trysail. However, I attempted rolling up the mainsail first.
This is a very awkward job with the sail lowered and the yacht
rolling violently. If the main-sheet is eased, the boom sweeps
from side to side in a devastating manner. I put various
lashings round boom and gaff, and hoisted the latter a shade.
After a two-hour struggle I got the sail more or less rolled up,
though it was impossible to get the turns evenly on the boom,
and I feared that the sail would be stretched out of shape if it
had to be hoisted in a gale. The wind moderated, and during

the night backed to north-east. I continued all day sailing comfortably to the WNW, and it was very pleasant to see the governor wheel of the patent log spinning rapidly instead of merely idling.

The thought of icebergs was very much in my mind, as the Ocean Passage Book states that they may be encountered within 500 miles of Newfoundland. There was also the possibility of field ice, which really would be rather awkward for a ship like mine. I remembered the loss of the *Titanic* very well. At the same time I was very loath to waste any of a fair wind. The seas are wide, and if no ice were seen during the day the risk of actually running into a berg during the short hours of darkness was slight and could be accepted.

At 2.45 a.m. on the 14th I woke to find that the ship had gybed and had hove herself to on the other tack. The wind had gone light, but it was rather disturbing to find that I was sleeping so soundly that the actual gybe did not wake me. The weather was getting very cold, and my log notes that I was wearing three jerseys, two pairs of trousers, and extra stockings. I had evidently reached the cold Labrador current.

During the afternoon the St John's broadcast came through in full loud-speaker strength. The advertisements were most interesting, and I felt that I was beginning to know this town with its Water Street, General Post Office, and shops which had just received a consignment of the latest Paris fashions.

Minor renewals, including a jib outhaul and a staysail tack lashing, were needed this day. So far nothing of importance has carried away, but the spinnaker gear is not heavy enough for prolonged use. Fog set in at dusk and the wind died away, leaving the yacht to drift through the night; but at least I could sleep soundly without the fear of running into ice.

The calm continued till noon next day, and the run was only twenty-five miles; but I was glad to get sights, as there had been none for four days. My latitude was only six miles out, but it was disappointing to find by an afternoon sight that the ship was twenty-seven miles farther away from my port than had been reckoned. I had been unable to pick up any definite time signal, and the error of the watch was dependent on an announcement from St John's that it was now three and a half minutes past two by the electric clock, Newfoundland daylight saving time. The Sailing Directions stated that the standard time of Newfoundland was 2 hours 31 minutes slow on Greenwich; presumably their summer time was one hour in advance, but it all seemed rather vague.

A fresh easterly breeze sprang up after lunch, and I felt glad to be sailing again. For supper a ham was boiled, which, with onions and potatoes, afforded a most palatable meal. It had been foggy for most of the day.

At midnight sail had to be shortened. The damp ropes were incredibly cold, and after a few minutes' work my fingers were numbed; but the sea was fairly smooth, so that it was dry in the cockpit.

I continued at the helm all night. Once a breaking crest showed white and gave me a start, thinking it was ice. The morning lightened imperceptibly, showing a dull, leaden sky and grim, grey seas. It was really very cold, and I piled on clothes, but was still unable to keep warm. I had no gloves, so pushed my hands into a pair of thick socks. Two stoves had been burning in the cabin throughout the night.

With the full daylight I fancied that there was a change in the formation of the waves difficult to describe, but perhaps the waves were smaller and steeper. The banks should have been fifty miles away, so that it was not likely to be due to shallower water. The mist continued; while drowsing at the helm a dirty white object caught my eyes. It was just under the lee bow, but before the helm could be touched the yacht had sailed over it. My heart seemed to come up to my mouth. I stood up trembling, but could see nothing either ahead or astern, and there had been no jar or grating of the keel. Then a porpoise provided the explanation as it showed its white-coloured belly.

June 17th was another day of calm and also of fog. Now that my port was not far off, the enforced delay and inaction was very irksome. For days I had seemed to be living in a little world of my own only about 500 yards across. During the morning my sense of isolation was suddenly broken by the sound of a steamer's syren. I jumped below to get my fog-horn, and replied with three short blasts, the signal for a ship running before the wind, though in fact the yacht had barely steerage-way. It is not likely that the puny squeak of my instrument was heard on board the steamer. For a time the noise of her syren increased; then it gradually died away, and my world contracted to its original isolation. The steamer never came into view.

The damp fog was penetrating every corner of the ship, and even the blankets were getting damp. I rigged clothes lines across the cabin and kept a stove burning continually. Only about half the stock of paraffin had been expended, so that no economy was needed. During the afternoon there was sufficient wind to sail, and I kept watch till midnight. The ship was then hove-to so that I could rest, as it seemed wise that I should husband my strength. If field ice should be encountered, it might be necessary to sail the ship and to keep on the alert for a prolonged period.

After four hours below I was eager to be getting on. The wind continued fresh from ENE., with the weather still thick and bitterly cold. I was quite unable to bear sitting in the cockpit and steered from below, taking a look round the horizon every few minutes. Having seen no ice, I was beginning to think that the bergs were merely a figment of the unduly pessimistic Sailing Directions, though as a matter of fact I had seen ice in the Atlantic on previous voyages.

At 9.20 a.m. the fog lifted suddenly. Fine on the weather bow and four or five miles distant was a white shape. It gave me a great start, and I stood up, every muscle taut. So icebergs really do exist. As I approached it could be seen to consist of three pinnacles, two tapering up to thin, sharp spires and the centre one truncated. It looked an enormous mass, but I hesitate to give an estimate in figures. The Sailing Directions recommend passing to windward to avoid small lumps of ice floating to leeward. Presently I went below to prepare my camera. Returning on deck I glanced ahead, but could see no berg. I had noted its bearing and looked at the compass for its exact direction. There, perhaps a half, or perhaps a quarter of a mile distant was just perceptible for a few seconds a ghastly white shape, half of it draped as if by a sheet, the fog. The curtain drew across, hiding it entirely. Single-handed in mid-Atlantic is no occasion to go seeking icebergs in a fog, so I hauled a couple of points to the north and proceeded on my way. Having noted the bearing of the berg, I could continue with the assurance of avoiding it, but how many more were there? Certainly for the hour or two during which it had been clear only this one had been in sight, and I rigidly kept in my mind that the mathematical chances of hitting one must be small.

At noon there is only another 100 miles to go. With the wind backing to NNE. the yacht steers herself very steadily with the wind abeam. Every four hours I read the log and reckon how many miles are left. What to do about making the land in this thick fog and possibly a strong wind, I leave for decision until the time comes. It is something of a fence to cross, but I will wait till I come to it, and from a study of the Newfoundland Sailing Directions I have great hopes that the northerly wind will clear away the fog. This occurred at 4 p.m. An hour later I made out several more bergs, one high and the others flat. They lay some miles to the north. The sun appeared for a few moments, and I got a snap at it. The ship was twenty-seven miles astern of the reckoning, but I was none too sure of the time. I have had no sight for latitude for several days, and may be many miles out.

At 8.45 p.m. I hoisted the working staysail in place of the big one, since with dusk coming on it was advisable to have the ship under instant command. The sea was comparatively smooth, so that I did not get wet while working on the fore-castle. For a couple of hours longer I remained at the helm, continually peering out into the dark ahead for possible icebergs; then, finding this strict lookout too much of a strain, I hove-to under very easy canvas. It was very snug and peaceful in the cabin, and the yacht made hardly any way, the governor wheel of the log not even turning.

Next morning I continued sailing at 5.30 a.m., wind fresh from the north and colder than ever; but I have used up all possible adjectives about the temperature. There is the usual note in my log about the spinnaker boom. This time the bracket aloft had carried away, and the topping lift had chafed through again. I got the spar down on deck, where it could be carried safely, though it is rather in the way.

I had reckoned that I might sight land about noon, and was keeping no very special lookout except for a casual glance ahead now and then for possible icebergs. I cooked a substantial breakfast, ate it, and cleared up in a quite leisurely manner. At 9 a.m. I came on deck and glanced ahead.

LAND! It seemed too good to be true, and I could hardly believe my eyes; but there it was, quite definite and unmistakable. A high and rocky coast-line lay ten to fifteen miles ahead with a conspicuous headland running to the north. Never again will it be possible to experience such a thrill as I did then as I gazed at the new continent spread out before me. In my excitement I stood up and patted the rail of the cockpit, shouting endearing names to my ship which had brought me so safely and surely across the ocean to America. Really there, certain and sure, and only a few miles ahead is the new continent, and I have succeeded. Hats off to Red Eric, Columbus, Cabot, and the rest; but this lonely run of mine, with day after day and week after week of nothing but sea around me, made me realize as never before the staunchness and drive of those early explorers, who did not know how far they had to go nor what they had to meet. Only one who had sailed for weeks on end without sighting anything can realize the immensity of the oceans and their loneliness.

But where was I? Obviously that distant shore must be Newfoundland, and that north-running cape is very definite. It should be Cape St Francis, but might possibly be Bonavista or some even more northerly headland. There was nothing to the southward that could possibly fit in with it. My last latitude was four days ago, and a hasty shot with a poor horizon at that. But don't panic or get too excited; previous landfalls have always been reasonably correct, and so probably is this one. I pored over the brand-new charts and feverishly read the coast description in the Sailing Directions. It all fits in with Cape St Francis, the point I was aiming at, and again almost too good to be true.

I steered for the headland, gradually becoming more and more certain that it was the right one. Conditions were ideal: a fresh breeze, clear weather, a clean bold coast and my harbour, easy of entrance, well to leeward. Surely now there could be no slip between cup and lip. As I approached the land, the detail of the coastline could be made out; to the southward two prominent headlands showed up which fitted in well with the picture on the chart as Redcliffe Head and the Sugar Loaf. When I was about five miles offshore I bore up confidently to the southward for the entrance to St John's, then about fifteen miles distant.

But now, lest I should get above myself, my pride had something of a fall. I sailed confidently on until I came to what appeared to be St John's Bay. When some half mile from the cliffs I felt suspicious. I could see no harbour entrance nor the lighthouse on Cape Spear, although all the bearings seem to fit in. Something was wrong; I tacked and hove-to, heading seaward while I pored over the chart again. I was in Tor Bay, and St John's lay six miles farther on. *Emanuel* continued her progress along the coast; to seaward was a small berg, but I was much too impatient to go out of my way for any photographs now. Soon I could make out the lighthouse on the real Cape Spear, and the harbour entrance lay before me clear beyond all doubt. There was a narrow gut between high cliffs, and the wind was going light, so that there might be difficulty in handling the ship under the land. I hove-to again and shook out all the reefs; next, the anchor was got up from below and shackled on to the cable, and I headed for the harbour once more. No pilot came out, nor was one needed, and I sailed into the Narrows. The wind dropped as I had expected, but before the yacht had lost her way came puffs, first on one bow and then on the other. The sheets were trimmed in a moment, and there was ample room for a handy little craft like mine to beat through the Narrows. The harbour then lay before me, with the town to starboard and rugged hill to port. Alongside one of the nearer wharves lay a large cruiser (H.M.S. *Dragon*), while ahead were a number of schooners at anchor. Hailing one, my first words for twenty-four days, I inquired where to anchor, and was told to let go anywhere. Rounding into the wind I let go and was soon surrounded by a crowd of boats from the schooners. Willing hands stowed my sails and launched the dinghy, and a few minutes later I was towed alongside Messrs Baird's wharf.

Subsequent worked-out positions on the chart showed that when land was sighted, the ship was sixteen miles north and three miles west of the dead-reckoning positions.

3
Portland Race

Where strong tides are diverted from their even courses by promontories or ledges, waves build up into alarming and dangerous shapes. These tide rips, races or overfalls are places to be avoided by yachts. Their positions are well-known and are marked on the charts, so the prudent sailor gives them a wide berth. Portland Bill, Colve Ohracium and Pentland Firth all have races which are hazardous at certain states of tide and wind. Those who get caught in a tide race can seldom find words to describe the experience. Hilaire Belloc's account will explain why.

A. R.

from '*THE CRUISE OF THE NONA*'
by HILAIRE BELLOC

AND so for rounding the Bill once more – a thing by me always dreaded – for Portland Race is a terrible affair.

Portland Race should, by rights, be the most famous thing in all the seas of the world. I will tell you why. It is a dreadful, unexpected, enormous, unique business, set right upon the highway of all our travel: it is the marvel of our seas. And yet it has no fame. There is not a tired man writing with a pencil at top speed in the middle of the night to the shaking of machinery in Fleet Street, who will not use the word 'Maelstrom' or 'Charybdis'. I have seen Charybdis – piffling little thing; I have not seen the Maelstrom, but I have talked to men who told me they had seen it. But Portland Race could eat either of them and not know it had had breakfast.

I have nearly always been successful in catching the smooth, narrow belt near the point: but I also once found that smooth belt fail me, and so have gone through the tail-end of Portland

Race; only the very tail-end. I had seen it from close by half a
dozen times before in my life. I had very nearly got into it
twice. But this time I did actually make knowledge of the 'thing
in itself' – and there is no mistaking it. It is one of the wonderful
works of God.

Portland Bill stands right out into the Channel and challenges
the Atlantic tide. It is a gatepost, with Alderney and the Hogue
for gateposts on the other side. They make the gate of the
narrow sea; outside them you are really (in spite of names) in
the air of the ocean, and look towards the Americas. Inside
them is the domestic pond. Portland Bill thrusts out into the
Channel and challenges the Atlantic sea. In my folly, for many
years I used to call it 'William', but I will do so no longer. For
there is something awful about the snake-like descending
point, and dreadful menace in the waters beyond. And here
Portland Bill differs from his namesake of the land. It would be
familiar to call a William of the land 'Bill'. But in the matter of
Portland it is familiar to call the Bill 'William'. I will never call
him William again.

I thought I had well known what the Race was before I
first heard it bellowing years ago. I knew after the fashion of
our shadowy nominal knowledge. I knew it in printed letters.
I knew it on the chart. I knew it in the Channel Pilot. Then I
came to it in the flesh, and I knew it by the senses, I saw it with
my eyes, and I had heard it with my ears. I had heard it roaring
like a herd, or park, or pride, of lions miles away. I had seen its
abominable waste of white water on a calm day: shaving it by a
couple of hundred yards. But there is all the difference in the
world between that kind of knowledge and knowledge from
within!

He that shall go through even the tail-end of Portland Race
in a small boat and in calm weather will know what he is
talking about, and for so vast an accession of real knowledge,
even that pain is worth while.

Portland Race lies in a great oval, sometimes three, sometimes four or five miles out from Portland Bill, like a huge

pendant hanging from the tip of a demon's ear. It is greater or smaller, according to whether the wind be off-shore or on, but it is immense always, for it is two miles or more across. It lumps, hops, seethes and bubbles, just like water boiling over the fire, but the jumps are here in feet, and the drops are tons.

There is no set of the sea in Portland Race: no run and sway: no regular assault. It is a chaos of pyramidical waters leaping up suddenly without calculation, or rule of advance. It is not a charge, but a scrimmage; a wrestling bout; but a wrestling bout of a thousand against one. It purposely raises a clamour to shake its adversary's soul, wherein it most resembles a gigantic pack of fighting dogs, for it snarls, howls, yells, and all this most terrifically. Its purpose is to kill, and to kill with a savage pride.

And all these things you find out if you get mixed up in it on a very small boat.

Perhaps the reason why Portland Race does not take the beetling place it should in the literature of England is that those who turn out the literature of England by the acre today never go through it, save in craft as big as towns – liners and the rest. Even these have been taught respect. During the War Portland Race sank a ship of 14,000 tons, loaded with machinery, and if you were to make a list of all the things which Portland Race has swallowed up it would rival Orcus. Portland Race is the master terror of our world.

And here I can imagine any man who had sailed saying to me that there are many other races abominable in their various degrees. I have not been through Alderney Race since the nineties, but I suppose it is still going strong. The Wild Goose Race you have already heard of – a very considerable thing. The Skerries also – I mean the one off Anglesey – is worthy to be saluted. And even little St Alban's, though it is a toy compared with Portland, is a nuisance in any wind.

But the reason Portland deserves the master name, which it has never achieved, the reason I write so strongly of the ignorance of England towards this chief English thing, the reason that Portland Race makes me seriously consider whether literary gents be not, after all, the guardians of greatness, and whether their neglect be not, after all, the doom of the neglected, is that this incredible thing lies to everybody's hand, and yet has no place in the English mind. The Saxon and Danish pirates of the Dark Ages must have gone through it (and – please God – foundered). Every one making Dorset from France for 2,000 years must have risked it. Today the straight course of innumerable ships out of Southampton, making for the Start and the Lizard to the ocean, leads them right past it – yet I know nothing of it in our Letters, unless it be one allusion of Mr Hardy's to the ghosts which wander above it. But there is no ghost so full of beef as to wander above Portland Race!

It is, perhaps, in that word 'Southampton' that I have struck the cause. Until Southampton became the port for the Americas the Race lay off the track. No man running down-channel from the Thames need touch the Race: no man running up. Even beating down or up-channel you are free to go about before you touch the broken water; running, you need not go near it. And steam need have nothing to do with it – except all that steam, which, during the last thirty years, has begun to use again more and more our one inland water of the south inside the Isle of Wight.

There is a great deal more I had intended to write about Portland Race. I had intended to talk about the folly of the Bill challenging the sea, and how it ought to be an island, as it was for centuries. I had intended to say something of that canal between Portland Roads and the West Bay, which ought to have been dug long ago, and which some day people will wish they had dug, when it is too late. I had intended to give rules for getting round by the narrow smooth. I had intended to curse the absurd arrangement whereby the tide, instead of behaving like a reasonable human tide, and running six hours either way, runs southerly nine hours out of the twelve from both sides of the Bay, leaving only three for the dodge round. I had intended to add much more.

But I cannot. Let me end with this piece of advice.

Never trust any man unless he has gone round Portland Bill in something under ten tons. Never allow any man to occupy any position of import to the State until he has gone round Portland Bill under his own sail in something under ten tons. But most of all, never believe any man – no, not even if you see it printed on this page – who says that he himself has done the thing.

4
Mediterranean Shipwreck

Most visitors go to the Mediterranean in high summer and find there calm seas and warm winds, so it comes as a shock when I hear of a disaster to a ship in those waters. Storms in the Mediterranean rise suddenly and can be very fierce. It happens that one of the earliest descriptions of a storm in the English language comes from this sea. I believe that St Paul was no sailor, but he noted carefully all that sailors did to save their ship and crew. However, he played one very important part in the final stages by raising their moral when he told them how God had revealed to him that the life of no one would be lost. To preserve good order and discipline up to the last minute before a ship finally goes down is vital to the saving of life.

A. R.

from '*THE ACTS OF THE APOSTLES*'
by ST PAUL

AND when it was determined that we should sail into Italy, they delivered Paul and certain other prisoners unto one named Julius, a centurion of Augustus' band. And entering into a ship of Adramyttium, we launched, meaning to sail by the coasts of Asia, one Aristarchus, a Macedonian of Thessalonica, being with us.

And the next day we touched at Sidon, and Julius courteously entreated Paul, and gave him liberty to go unto his friends to refresh himself. And when we had launched from thence, we sailed under Cyprus, because the winds were contrary. And when we had sailed over the sea of Cilicia and Pamphylia, we came to Myra, a city of Lycia. And there the centurion found a ship of Alexandria sailing into Italy, and he put us therein.

And when we had sailed slowly many days, and scarce were come over against Cnidus, the wind not suffering us, we sailed under Crete, over against Salmone. And, hardly passing it, came unto a place which is called The Fair Havens, nigh whereunto was the city of Lasea.

Now when much time was spent, and when sailing was now dangerous, because the fast was now already past, Paul admonished them, and said unto them:

Sirs, I perceive that this voyage will be with hurt and much damage, not only of the lading and ship, but also of our lives.

Nevertheless the centurion believed the master and the owner of the ship more than those things which were spoken by Paul, and because the haven was not commodious to winter in, the more part advised to depart thence also, if by any means they might attain to Phenice, and there to winter, which is an haven of Crete and lieth toward the south-west and north-west. And when the south wind blew softly, supposing that they had obtained their purpose, loosing thence, they sailed close by Crete.

But not long after there arose against it a tempestuous wind called Euroclydon; and when the ship was caught, and could not bear up into the wind, we let her drive. And running under a certain island which is called Clauda, we had much work to come by the boat, which when they had taken up, they used helps, undergirding the ship; and fearing lest they should fall into the quicksands, strake sail, and so were driven.

And we being exceedingly tossed with a tempest, the next day they lightened the ship. And the third day we cast out with our own hands the tackling of the ship; and when neither sun nor stars in many days appeared, and no small tempest lay on us, all hope that we should be saved was then taken away.

But after long abstinence Paul stood forth in the midst of them, and said,

Sirs, ye should have hearkened unto me, and not have loosed from Crete, and to have gained this harm and loss; and now I exhort you to be of good cheer, for there shall be no loss of any man's life among you, but of the ship. For there stood by me this night the angel of God, whose I am, and whom I serve, saying, Fear not, Paul; thou must be brought before Caesar: and lo, God hath given thee all them that sail with thee. Wherefore, sirs, be of good cheer, for I believe God, that it shall be even as it was told me. Howbeit we must be cast upon a certain island.

But when the fourteenth night was come, as we were driven up and down in Adria, about midnight the shipmen deemed that they drew near to some country; and sounded and found it twenty fathoms. And when they had gone a little farther, they sounded again, and found it fifteen fathoms. Then, fearing lest we should have fallen upon rocks, they cast four anchors out of the stern, and wished for the day.

And as the shipmen were about to flee out of the ship, when they had let down the boat into the sea, under colour as though they would have cast anchors out of the foreship, Paul said to the centurion and to the soldiers,

Except these abide in the ship, ye cannot be saved.

Then the soldiers cut off the ropes of the boat and let her fall off.

And while the day was coming on, Paul besought them all to take meat, saying,

This day is the fourteenth day that ye have tarried and continued fasting, having taken nothing. Wherefore I pray you to take some meat, for this is for your health; for there shall not an hair fall from the head of any of you.

And when he had thus spoken, he took bread and gave thanks to God in presence of them all, and when he had broken it he began to eat. Then they were all of good cheer, and they also took some meat. And we were in all in the ship two hundred threescore and sixteen souls. And when they had eaten enough, they lightened the ship, and cast out the wheat into the sea.

And when it was day, they knew not the land, but they discovered a certain creek with a shore, into the which they were minded, if it were possible, to thrust in the ship. And when they had taken up the anchors, they committed themselves unto the sea, and loosed the rudder bands, and hoisted up the mainsail to the wind, and made toward shore. And falling into a place where two seas met, they ran the ship aground, and the fore-part stuck fast and remained unmovable, but the hinder part was broken with the violence of the waves.

And the soldiers' counsel was to kill the prisoners lest any of them should swim out and escape. But the centurion, willing to save Paul, kept them from their purpose, and commanded that they which could swim should cast themselves first into the sea, and get to land, and the rest, some on boards, and some on broken pieces of the ship. And so it came to pass, that they escaped all safe to land.

5
Trade Winds and Doldrums

Trade winds which blow with reasonable constancy both in force and direction are of great benefit to sailors. Just as the Trades are the most popular, then the Doldrums between the two great Trade-wind belts must be the most unpopular conditions for sailing. Very few people have written about the Doldrums, or any other calms for that matter. I suppose sailors are so irritated by the swinging of the limp sails and the almost lifeless motion of the boat that they do not feel inclined to sit down and write about them. Authors restrict themselves to a brief description of the conditions, express their exasperation and put down their pens until the wind comes.

Unfortunately, I have been unable to discover anything about Rex Clements except what he tells us in his book, and even there I can find no dates. But he sailed as an apprentice on the British three-masted barque *Arethusa* on a voyage round the world. Six years in all, as apprentice and officer, he served in her, and then passed on to other rigs, finally settling staidly into steam. And as for the *Arethusa* – he tells us sadly that one summer's day in 1917, when homeward bound and off the Fastnet, she encountered a German submarine. It was the end, and she went down gloriously, with all sails set, the red ensign at her peak and an old shark's tail pointing bravely to the sky.

A. R.

from '*A GIPSY OF THE HORN*' by REX CLEMENTS

AS we were now drawing near the region of the steady trade winds and the weather had become quite settled, the opportunity was taken to shift sail. A ship, unlike the majority of her sex ashore, always wears her best clothes in bad weather, and when skies are soft and winds gentle changes into her oldest and most worn attire. Appearances have nothing to do with the choice; it is a matter that the winds decide, for only the newest and strongest canvas can sustain the winter gales of the North

Atlantic or the world-encircling sweep of the great winds of the
Southern Ocean.

Early, then, one morning all hands were roused out to make
the necessary change. The suit it was proposed to bend was
brought on deck, and gantlines were rove at the mastheads in
readiness. As four bells struck all hands tumbled out, the order
was given 'Port watch for'ard; starboard to the main', and the
work began.

First came the two courses. These were unbent, lowered on
deck, rolled up and put away. Those that were to take their
places were stretched out, bent on to the buntlines and hoisted
aloft, where they were secured to the jackstay and, the sheets
being shackled on and hauled aft, speedily set.

After the courses and a break for breakfast came the lower
tops'ls – those aloft unbent, the fresh ones bent and promptly
sheeted home. Following these, the upper tops'ls and then the
t'gallants'ls. The work early developed, as no doubt the old
man intended it should, into keen rivalry between the watches
and an eager race as to which should finish their mast first.
By the time the t'gallants'ls were being swayed aloft all hands
were working schooner-rigged, going at it with their blood up.
The royals went aloft to the hand-over-hand shanty:

> 'Way-ay! and up she rises,
> Way-ay! and up she rises,
> Way-ay! and up she rises,
> Early in the morning.'

The word 'morning' was hardly reached before the sails were
snatched out along the yard, the sheets whipped on and a
triumphant hail 'All ready, hoist awa-a-ay!' sent ringing to the
deck.

It was a well-fought fight and at the end of the eight hours it
lasted was only won by the Mate's watch by a matter of
minutes – thanks largely to the bull-strength and the fiery
activity of Paddy.

After a belated dinner the Second Mate's men tackled the

mizzen, while we of the port watch went for'ard to the jibs and thereafter, by twos and threes, to the stays'ls. It had turned three bells in the first dog watch before the last sail was bent and set, the old suit bundled below into the locker and all hands trooped off to their tea.

Beyond a call to relieve the wheel, no order came from aft till eight bells were struck and we mustered for the roll-call and the routine of the night watch. There was no sing-song in the half-deck that evening though, we were too tired to do more than smoke our pipes and contentedly contemplate the day's work.

All this time the fresh breeze blew and slowly hauled into the nor'-east. On the 24th of January we entered the tropics, with the wind a steady whole-sail breeze and the horizon piled with fleecy masses of white cloud. There was little doubt that we had got the Trades at last. Two days later when we were in the latitude of the Cape de Verde islands, having run 500 miles in the forty-eight hours, there wasn't a doubt of it. The steady, singing breeze, true to a handsbreadth; the tumbled whites and blues of the sea all a-sparkle in the sunshine; the untroubled rim of the far horizon – these were signs unmistakable of the authentic Trades, the glorious winds that make seafaring a pleasure cruise.

No need to touch tack or sheet now; the ship sailed herself and the helmsman had an easy time, a spoke or two now and then was all she needed. The days began to be a sheer delight. Most of the time we were employed aloft – serving, parcelling, repairing, renewing, attending to the hundred and one jobs that a sailing-ship is constantly in need of and that suggest comparison of her with a lady's watch – always out of repair. There is nothing more delightful imaginable than sitting astride a gently-swaying yard, high above ship and ocean, with a pot of tar and a ball of spunyarn, for long hours on end, casting an occasional glance at the acre of foam under the bows or the long white wake astern that tells of a good passage to be made.

Running the Trades down in a sailing-ship is an experience worth having – a taste of sea-life at its best. No words can well convey the beauty, the freedom, the glorious exhilaration of it all. The long halcyon hours wing by, with blue sea beneath and blue sky above and unending glitter of spray and sunshine. The ocean is all one's own or shared with a white seabird or two. The day seems twice as long as usual, and when the western sky floods with rainbow hues and the day's glory of gold turns to the silver splendour of night, there comes never an abatement or alteration in the steady onrush of the wind.

It would be hard to find a more beautiful sight than that of a sailing-ship running through the Trades as she appears, when viewed on a moonlit night, from the end of her own bowsprit.

Perched out there one seems alone in space, projected into a world of emptiness and utter beauty. Ahead the shoreless sea stretches dimly out into a mysterious eternity. The dark vault of the sky is powdered with stars, save where the sailing moon rides high and pours on the quiet sea a river of white radiance, a pathway to unimaginable realms of faerie.

And how lovely the swaying vision of the ship herself! In grace of outline and harmony with sea and sky she seems none other than the winged Spirit of the Night. Beneath one's perch the sharp stem shears unendingly on through the dark tumbling water, smashing it into a splinter of white foam; while aloft, rising tier on tier in contours of unmatchable symmetry, a mighty cone of canvas, carved in rigid ivory or blackest ebony, sways silently amid the stars.

'There be triple ways to take, of the eagle or the snake,
     Or the way of a man with a maid;
But the sweetest way to me is a ship's upon the sea
     In the heel of the North East Trade.'

With the coming of the Trades we apprentices had more time to look about us and we began to take an interest in the sea itself. Frequently in the dog watch the old man called one or other of us up on the poop and talked to us about our profession and duties. On such occasions he always used to emphasize the importance of learning to know the face of the sky and sea and studying the ways of birds and fishes. He himself was no great reader, but his knowledge of winds, weather and natural history was truly wonderful. He navigated the ship more, I verily believe, by the look of the sea, the flight of birds, the appearance of clouds and a certain indefinable sixth sense emanating from his vast knowledge of all these, than by the more orthodox method of scientific instruments. I never grasped half what he told me about such things as the movements of upper and lower clouds and the behaviour of birds, but he was extraordinarily interesting to listen to and talked to us 'boys' in a very kind and fatherly way when there was no work to be done. When there was, we had to jump like steel wires.

Much of his weather-wisdom was transmitted to us in the form of pithy deep-sea sayings – proverbs and scraps of doggerel easily memorized. 'The devil would make a sailor if he'd only look aloft' was a favourite aphorism of his, emphasizing the importance of keeping one's weather eye lifting. Another, and one that we often used to repeat in squally weather, was:

> 'With the rain before the wind
> Your tops'l halyards you must mind:
> With the wind before the rain
> Your tops'ls you may hoist again.'

'Look out when you see a fan-shaped squall'; 'A clear eye to the south'ard is the sure sign of a wet jacket' and 'The sun setting up the backstays – a deep-sea synonym for 'sun-dogs' – means rain,' were others of his sayings. Very true, too, we found the verse:

> 'When the sun sets behind a bank
> A westerly wind you shall not want:
> When the sun sets clear as a bell
> An easterly wind as sure as hell.'

while I always considered the warning, 'Any fool can make sail, but it needs a good sailor to take it in,' a priceless scrap of wisdom.

And philosophy too, in the same concentrated form, we imbibed from our commander, as well as from the men for'ard. What could be sounder than 'Obey orders, if you break owners' – meaning, do as you're told, even if you know it's wrong – or more wisely admonishing than 'Growl you may, but go you must'. 'One hand for yourself and one for the owners,' too, is sound advice to remember when aloft, and 'More days, more dollars' a comforting reflection on a long passage. 'He who would go to sea for pleasure would go to hell for a pastime' is an attempt at heavy satire, as the saying, 'To work hard, to live hard, to die hard and then to go to hell would be hard indeed,' marks an effort at consolation.

Captain West was a typical seaman of the old school – rough, masterful and equal to all emergencies. Steamers he held in a large-hearted contempt, but the sea he knew and loved. His handling of the ship in heavy weather or narrow waters was an education to watch, and all the time I sailed with him I never knew him make a bad landfall, or be out in his rough and ready reckoning more than a handsbreadth from the truth. Some of his methods would give a modern liner-officer cold shudders. Once when I was taking an azimuth he swept me and the instrument aside with a lordly gesture. 'A handful of degrees don't matter, my boy,' he said, 'take the sun like this' – laying a great hand edgeways on the compass and squinting along it at the sun – 'and keep your eyes about you.'

In one way or another our commander played a large part in our lives. All the time I knew him his was always the dominant personality on board. And he influenced very considerably the apprentices who sailed under his command. Years after, in Cape Town, I met the chief officer of a large liner – the acme of modern smartness and efficiency – who, when I mentioned Captain West, said he had served his time under him and often congratulated himself on the fact. The skipper was a big man in every way, in personality as in physique; one might hate him or one might like him, but it was impossible to ignore him. From the Mate downwards we all stood more than a little in awe of him, but it was an awe mixed with liking, and when other ships and other men were in company we took pride in telling of his skill and masterfulness.

But to hark back to the Trades. One of the first denizens of deep water, and one of the most interesting, that we saw as soon as we got down to warmer latitudes was flying-fish. Just a few at first, then shoals and shoals of them. They simply swarm all over the tropic seas, and though they are preyed upon by numberless enemies – bonito, coryphene and albicore – their numbers are almost infinite. 'Flying-fish weather,' sailors often call the latitudes they inhabit. Their so-called wings are thin gauzy affairs, bearing no resemblance to those of a bird but more like a huge dragonfly's. Their flight too is really no more than skimming; they rise out of the water with a flip of their tails and vibrate their wings like planes. A couple of hundred yards is about their limit, then back they fall into the water with a 'zip' and, a moment after, leap out again and are off on another flight. I have often heard it said that they can fly only as long as their wings are wet; whether this is so I don't know, but it is certain that they only remain in the air ten or twenty seconds and can do little more, I should think, than throw their pursuers momentarily off their track.

They keep together in huge shoals and sometimes we saw an acre of more of the surface break into a silver shimmer as thousands upon thousands of them rose into the air, skimmed madly along just above the tops of the seas, and suddenly disappeared again. Day and night seem just the same to them, they are for ever a-twitter and on the alert. In the mornings we often picked up one or two in the scuppers near the side-lights, that had fallen aboard in the dark. They seem to be attracted by a light, so we tried the old dodge of displaying a gleam from our lamp overside at night, carefully screening it, of course, from the officer on watch, and were often rewarded by having a plump flying-fish falling on board with a smack, all ready to cook for breakfast.

In size and appearance the flying-fish is somewhat like a

herring and in taste not dissimilar. Most of those we caught had two wings, but some had four, and a monstrous fellow we caught far down south had three complete pairs of wings and was nearly eighteen inches long.

Another and a very different species of deep-water-man that we met with were whales. I well remember the first that I saw. It was one evening, off the Canaries. I was looking out over the rail, waiting for two bells to strike and tea to be served out; when I saw, quite close to the ship, one – two – three – up to half a score, huge black bodies heave themselves silently half out of the sea. As they rose, thin jets of watery vapour were thrown high in the air with a long hissing noise. 'Whales, by all that's wonderful!' said I to myself. They seemed in no hurry, but with an easy gliding roll that hardly rippled the water went curving under. A moment or two later up they came again – one saw the glistening curve of each huge bulk – then down once more they plunged. They were most fascinating to watch, in movements so vast and leisurely, yet they did not ruffle the surface as much as a swimming man would have done.

One Sunday morning, a week or two later, I had an even better opportunity of observing a whale closely, for an immense fellow came up alongside, almost touching our plates, and swam for a few minutes abreast of us. I believe he was a solitary old bull-cachalot; he was without comparison the most montrous living thing that ever I saw. He stretched from the break of the poop almost to the fo'c'sle head and cannot have been far short of a hundred feet in length. His skin was a dirty grey in colour, not black and glistening like the majority of whales, and covered with barnacles and what looked like growths of weed. He swam for a few minutes alongside and then, with majestic deliberation, sounded.

A curious thing about whales is that their tails are not fore-and-aft like those of fishes, but set on thwartships with the flukes to port and starboard. We saw great numbers at different times and a whole host of varying species. Sperm whales were the biggest and humpbacks the most common, at least in warm latitudes.

These latter are amusing creatures to watch. They have a trick of rolling over and over with their flippers waving in the air as though they were immensely tickled at something and when they sound they do it wholeheartedly. They stand upright on their heads, with their lower halves rising straight out of the water, and go down with a rush, their tail-flukes wagging 'goodbye'. Like the whales of the polar seas, the humpbacks live on small surface crustaceans – called 'whalefeed' by sailors – that are met with floating about in huge fields, sometimes miles in extent. These fields, that 'oft bank the mid-sea', are composed of all sorts of little creatures, the most noticeable being a small shrimp-like thing that swims tail first and looks not unlike a tiny boiled lobster. It staggers one to think how many must be required to make a decent meal for a single whale.

Unlike all these species, the jaws of which are furnished with whalebone, the sperm whale or cachalot has a huge mouth armed with great teeth and feeds on squids and octopus. The sperm whale is the true monarch of the sea and 'mightiest that swims the ocean stream'. There is no more tremendous exhibition of strength than the sight of a cachalot 'breaching', that is, jumping clear out of the sea and falling back with a mighty 'thwack' that can be heard miles away.

Ever as we sagged south, the sun's rays increased in intensity, and, as we drew near the Equator, the good breeze began to fine away. It carried us to about the fifth parallel of North latitude and then vanished altogether. The midday sun flamed down on us almost straight overhead, the sea took on an oily sluggish appearance, and, after a few aimless and wandering gusts, the breeze died away and we were left becalmed – the centre of a clock calm. The Doldrums had us in their grip.

'There passed a weary time.' We lay helpless on an oily sea, the spars sticking up idly into the still air. The heat grew terrific; bare iron was too hot to touch and the pitch bubbled

out of the seams between the planking and stuck to our bare
feet. From the yards the sails hung listlessly in heavy folds;
we hauled the uselessly-flapping mains'l up, and listened to
the grind and rasp of the parrals as she rocked to some im-
perceptible underrunning swell.

It was exasperating. The man at the wheel leaned idly over
the spokes, for she had lost steerage way; the half-deck was like
a furnace and our only amusement was scanning the face of the
brazen heavens and whistling for a wind.

At times the sky would become overcast, a black thunder-
cloud overspreading everything and then, without a moment's
warning, down came the deluge! The clouds seemed to collapse, the rain fell in sheets and torrents, with such intensity
that it was impossible to see more than a few yards, and the sea
gave off a continuous hissing roar under the violence of the
downpour.

Then, as suddenly as it began, the rain stopped, the clouds
rolled away and the sun shone down again with undiminished
vehemence upon our drenched and steaming barque.

Sometimes at the approach of one of these squalls we stopped
up the scupper-holes, and in a few minutes the main deck was
inches deep in water. All hands would rush out, stark naked,
with soap and dirty clothes and indulge in an impromptu bath
and washing day for the quarter of an hour or so that the deluge
lasted.

The vagaries of the wind, too, were heartbreaking. For the
greater part of the time no breath of air stirred to quiver the
dull folds of canvas or cool our heated bodies. If a match were
struck its flame burnt steadily upward without a flicker. Then
the merest suggestion of a breeze would whisper from ahead
and round the heavy yards would be dragged to take advantage
of it. No sooner had they been braced up than, as if in mockery
of our efforts, a faint puff of air would come from the opposite
quarter. Savagely the officer would give the order 'Weather
fore brace!' and savagely the watch would throw down the
freshly-coiled gear and haul the yards round to the new tack.

So it went on, day and night; the ship making a bare mile of headway in the twenty-four hours and everybody's temper frayed to breaking-point.

But even in these regions hated of the sailor we had some little excitement, for several sharks appeared. The old man first noticed one lazily swimming under the counter and – keen sportsman that he was – sent me to fetch the shark hook. This was a fearsome-looking implement, fully a foot in length and sharply barbed and pointed; a swivel and several feet of chain were attached to it to avoid turns and prevent the shark severing it with his teeth.

Seeing what was afoot, Tommy came running along with a lump of pork. This was securely lashed to the hook with wire, and a coil of stout ratline-line was bent on to the chain. When all was ready the end was made fast and the old man threw the hook overboard.

The shark was then swimming on the far side of the counter. He seemed to scent food, for he turned in his tracks and swam leisurely towards the bait. As he drew abreast of it he half turned on his side. We saw the white gleam of his belly as, without an attempt at finesse, he made a swift snap at the pork. It disappeared, there was a violent tug on the line, and a shout from the old man. Burton and I dashed to bear a hand and several of the men, who were peeping expectantly over the poop ladder, ran up. The shark was soon mastered and dragged, jerking frantically, up to the rail and inboard. He was not a very big one, only about five feet in length, and a few heavy blows on the head with the carpenter's maul settled him.

Hardly was he dispatched before Burton gave a shout and pointed astern. Following the direction of his hand, we saw the triangular fin of another shark gliding through the water well out on the quarter. As quickly as we could we rebaited the hook and dropped it overboard again. For a few minutes it trailed there, seemingly without attracting the attention of the shark. Then suddenly the cruising fin disappeared, but before we had time to feel disappointed, we saw a monstrous blue shadow glide under the counter a few feet from the bait. It was a monster this time and no mistake.

The shark seemed to take no notice of the morsel we had provided for him. He swept past in a half-circle, swimming about six feet deep. Almost out of sight he went, but came back. He swam slowly around for a few minutes, then, giving a flick with his tail as though he were bound somewhere and it was time to forge ahead, he suddenly turned, made a swift lunge upward at the pork, and was hooked.

Half a dozen of us flung ourselves on the line and hauled for all we were worth. It was a brand-new, 21-strand hemp, but as we felt the strain, we thought it would never hold him. Fighting desperately, we managed to drag him just clear of the water and, as we did so, with a last furious convulsion he lashed himself free.

'Come up the line!' yelled the old man and, as we dropped it, flung the hook back into the water.

Then we had an example of the shark's voracity.

The hook had hardly touched the water before the great brute leapt at it again, biting so furiously that he almost swallowed the pork and the barb came through half-way down his throat. There was no doubt about it this time. Putting all our strength into it, we dragged his head well above water. He was too big a fellow to haul aboard in the same way as our previous capture, so a running bowline was sent down the line, shaken over his body and hauled taut close to his tail. The bight was then passed through a snatch-block made fast to the boom-end and we hauled him up horizontally. When high enough, he was swung inboard with lines to head and tail, dragged to the break of the poop and dropped down to the lower deck with a crash.

He was a good ten feet long and lashed about so furiously that he completely took charge. For a few minutes we couldn't get near him. Then Stedman managed to ram a windlass bar down his throat as the rest of us hung on to the lines that were fast to either end of him, and Chips, sliding a hatch under his after part, cut off his tail with an axe.

After that we set to work to cut both fish up. The tail of the bigger one was triumphantly carried for'ard and nailed to the extremity of the bowsprit, the longer lobe uppermost. This is always the custom when a bigger shark than usual is caught, and constitutes the time-honoured decoration of a deep-water-man. The backbone of the smaller shark the steward annexed to make into a walking-stick. When completed, this was a truly nautical affair: the vertebrae were cleaned and the holes filled in with red and black sealing wax. An iron rod was passed down its whole length to stiffen it and a complicated wall-and-crown knot woven on top to serve as a handle.

From the smaller one the men cut steaks to be cooked for their supper. I tried some, but thought it very dry and tasteless. The Mate cut some strips off the skin, which is exceedingly rough and dries as hard as a board, for cleaning purposes in place of sand-paper. He also had both the livers cut out that they might decompose and exude their oil. The tin in which these choice morsels were put was kept in the forepeak and rendered that part of the ship uninhabitable until the operation was concluded.

Next day we caught two more sharks, both about six feet long. The men called them shovel-nosed sharks and I cut out the jaw of one, with its triple rows of triangular knife-edged teeth, to keep as a curiosity.

For all their voracity sharks are lazy brutes. They are deservedly hated by most seafarers, but the South Sea islanders hold them in contempt. A Kanaka will slip into the water and tackle one with only a knife, meeting it head on, passing underneath and ripping up its belly as he passes. Even little boys will dive into shark-infested waters for the sake of a small coin thrown overboard. But these are Kanakas and pretty well amphibious. Most white men – sailors especially – detest sharks.

Their voracity is awful. On one occasion, in Port of Spain,

we threw overboard a dead pig. I watched it as it floated astern. Suddenly I saw the fin of a shark cut through the water towards it. There was a gleam and a splash and the body of the pig, streaming blood, was knocked flying. Instantly, appearing from nowhere, a dozen fins shore through the water close to the carcase. There was a violent commotion, the pig shot bodily into the air and hardly touched water before it was knocked clear again. For a few moments its carcase was bandied about in a swirling eddy of bloody water. Half a dozen times it soared upwards before the headlong rush of a blunt snout, while half a score lean blue bodies leapt and snapped about it. A couple of minutes the commotion lasted, then stillness fell, with just a tinge of blood lingering on the water. It was an object-lesson in ferocity and made one pray for a dry death.

A shark is nearly always accompanied by a companion, a graceful little chap called a pilot-fish. Something like a perch the latter is, with pretty barred markings. I never heard any reason advanced as to the connection between the two. Perhaps the pilot-fish searches for food for its big friend and, at any rate, seems itself to be safe from the other's omnivorous appetite. To the pilot-fish the advantage of the companionship is still more difficult to understand.

A shark is often attended by another satellite or two in the shape of a remora or sucking-fish. Unlike the pilot-fish the latter has no claim to good looks. It is about nine inches long and scaleless. On the upper part of its flat head is an oval disc, with soft transverse ridges and a cup-like rim, possessing a tremendous power of suction. By means of this disc the sucker attaches itself to the shark and is carried about without an effort on its own part. So tenaciously does it cling that a shark is often hauled aboard with the sucking-fish still fast to it. Laziness can hardly go further: the sucker is about the last word in that direction.

For a whole week we lay drifting about on a windless sea. Save for a few minutes at a time, when a catspaw ruffled the surface of the water, we were without steerage way. The excitement afforded by the sharks – we caught eight during the week – was a welcome relief to the wearisome monotony of hauling the creaking yards round to every puff of air.

One morning the skipper spied a dolphin gliding and darting through the water on the port side of the poop. He sent for the grains, a weapon like a double trident with five barbed points, fitted to a slender wooden haft and having a line attached. Then he climbed out on the bumpkin and, watching his opportunity, drove the grains down just as the dolphin flashed beneath him. He speared it the very first shot, a piece of skill worthy of our Nimrod of a skipper.

We quickly hauled the dolphin aboard and a glorious fish he was, nearly five feet in length, with scintillating metallic hues. He looked superb gliding through the water like a flash of many-coloured light, but his brilliancy soon dimmed as he lay quivering on our sun-scorched deck. We carried him for'ard and cut him up and cabin, fo'c'sle and half-deck had a sumptuous tea, for the flesh of the dolphin is white, dry and very good eating.

As far as I can gather there are two entirely different species of fish called by the name 'dolphin'. That just spoken of is the only one to which sailors give the name. It is a long, narrow fish, with a blunt, rounded head and a long fin running all down its back. Another name for it is the 'coryphene', and in South America it is called the 'dorado', or 'golden one', from its appearance in the water. It is the inveterate enemy of the flying-fish.

The other creature called a dolphin isn't really a fish at all, but a mammal and allied to the porpoise family, from which it differs in being slimmer and having a longer snout. This is the dolphin one so often sees depicted gambolling around the edges of seventeenth-century maps and books of voyages. The two species, though so unlike, are often confused. When, for instance, Byron says that 'Parting day dies like a dolphin' it is the first kind he is thinking of; but when he speaks of 'Sporting dolphins bending through the spray' it must be in allusion to the second.

The many kinds of fish we saw caused the old man to over-haul his gear, and very soon the poop was decorated with the implements of the deep-sea fisherman's craft. On one side, slung in beckets, was a heavy two-flanged harpoon fitted to an eight-foot shaft, with a thick line attached and almost strong enough to hold a whale. On the other side was a smaller har-poon with a single barb that drew across in the shape of a T, a deadly weapon of the type used by the Yankee whalemen of Nantucket and New Bedford. Underneath were the grains and coiled near the rail the shark hook and line. All were ready for instant use and the old man never let slip an opportunity of bringing them into action.

All these things helped to while away the tedium of the Doldrums. And then, at last, to our relief, wind came. It was at sunset on the eighth day of windless drift that we saw, well out on the port bow, a dark line on the water and felt the merest suggestion of movement stir the heavy air. The dark line drew closer and, as it did so, resolved itself into a ruffled patch on the surface of the greasy sea. The old man was watching it intently. A few moments satisfied him of what he saw, for, coming to the poop-rail, he gave the order 'Starboard fore brace' in a tone that betokened he anticipated a change of some sort.

The little ripples touched the ship's side and cool breath of air fanned our cheeks. At the order we sprang to the braces and hauled cheerily.

The heavy yards swung round, the sails flapped and filled and the barque began to make way through the water.

The welcome breeze kept on blowing gently but steadily. The main sheet was hauled aft and the hitherto useless stays'ls set. The gear was coiled down and still the gentle breeze blew and the cheerful sound of rippling water arose.

At eight bells the wind still held and the ship was logging an

easy five knots. As the wheel was relieved and we turned away
for'ard, Mac voiced the general opinion:

'Lay your shore-gear handy, me boys,' said he, 'the port
watch ha' brought ye the Trades.'

And so it proved; all night the breeze held, blowing steadily
from the sou'-sou'-east and imperceptibly increasing in force.
It was about 1° North latitude where we picked them up and
next morning, the 14th of February, word went round that we
should be crossing the Line that day.

6

The Pilot of the Pinta

A man who permanently worried about his health would surely never go to sea, at least not by himself, as a lone voyager has all his time cut out looking after his ship when he is well. But the sailor can sustain damage, as can his ship, or he can become ill and unable to manage the vessel. Many lone sailors have experienced this misfortune. Slocum's account of such an incident makes light of what must have been a very worrying time.

A. R.

from 'SAILING ALONE AROUND THE WORLD'
by JOSHUA SLOCUM

I SET sail from Horta early on July 24th. The south-west wind at the time was light, but squalls came up with the sun, and I was glad enough to get reefs in my sails before I had gone a mile. I had hardly set the mainsail, double-reefed, when a squall of wind down the mountains struck the sloop with such violence that I thought her mast would go. However, a quick helm brought her to the wind. As it was, one of the weather lanyards was carried away and the other was stranded. My tin basin, caught up by the wind, went flying across a French schoolship to leeward. It was more or less squally all day, sailing along under high land; but rounding close under a bluff, I found an opportunity to mend the lanyards broken in the squall. No sooner had I lowered my sails when a four-oared boat shot out from some gully in the rocks, with a customs officer on board, who thought he had come upon a smuggler. I had some difficulty in making him comprehend the true case. However, one of his crew, a sailorly chap, who understood how matters were, while we palavered jumped on board and rove off the new lanyards I had already prepared, and with a friendly hand helped me 'set up the rigging'. This incident gave the turn in my favour. My story was then clear to all. I have found this the way of the world. Let one be without a friend, and see what will happen!

Passing the island of Pico, after the rigging was mended, the *Spray* stretched across to leeward of the island of St Michael's, which she was up with early on the morning of July 26th, the wind blowing hard. Later in the day she passed the Prince of Monaco's fine steam-yacht bound to Fayal, where, on a previous voyage, the prince had slipped his cables to 'escape a reception' which the padres of the island wished to give him. Why he so dreaded the 'ovation' I could not make out. At Horta they did not know. Since reaching the islands I had lived most luxuriously on fresh bread, butter, vegetables, and fruits of all kinds. Plums seemed the most plentiful on the *Spray*, and these I ate without stint. I had also a Pico white cheese that General Manning, the American consul-general, had given me, which I supposed was to be eaten, and of this I partook with the plums. Alas! by night-time I was doubled up with cramps. The wind, which was already a smart breeze, was increasing somewhat, with a heavy sky to the sou'west. Reefs had been turned out, and I must turn them in again somehow. Between cramps I got the mainsail down, hauled out the earings as best I could, and tied away point by point, in the double reef. There being sea-room, I should, in strict prudence, have made all snug and gone down at once to my cabin. I am a careful man at sea, but this night, in the coming storm, I swayed up my sails, which, reefed though they were, were still too much in such heavy weather; and I saw to it that the sheets were securely belayed. In a word, I should have laid to, but did not. I gave her the double-reefed mainsail and the whole jib instead, and set her on her course. Then I went below, and threw myself upon the cabin floor in great pain. How long I lay there I could not tell, for I became delirious. When I came to, as I thought, from my swoon, I realized that the sloop was plunging into a heavy sea, and looking out of the companionway, to my amazement I saw a tall man at the helm. His rigid hand, grasping the spokes of the wheel, held them as in a vice. One may imagine my astonishment. His rig was that of a foreign sailor, and the large red cap he wore was cockbilled over his left ear, and all was set off with shaggy black whiskers. He would have been taken for a pirate in any part of the world. While I gazed upon his threatening aspect I forgot the storm, and wondered if he had come to cut my throat. This he seemed to divine. 'Señor,' said he, doffing his cap, 'I have come to do you no harm.' And a smile, the faintest in the world, but still a smile, played on his face, seemed not unkind when he spoke. 'I have come to do you no harm. I have sailed free,' he said, 'but was never worse than a *contrabandista*. I am one of Columbus's crew,' he continued. 'I am the pilot of the *Pinta* come to aid you. Lie quiet, señor captain,' he added, 'and I will guide your ship tonight. You have a *calentura*, but you will be all right tomorrow.' I thought what a very devil he was to carry sail. Again, as if he read my mind, he exclaimed: 'Yonder is the *Pinta* ahead; we must overtake her. Give her sail; give her sail! *Vale, vale, muy vale!*' Biting off a large quid of black twist, he said: 'You did wrong, captain, to mix cheese with plums. White cheese is never safe unless you know whence it comes. *Quien sabe*, it may have been from *leche de Capra* and becoming capricious . . .'

'Avast, there!' I cried. 'I have no mind for moralizing.'

I made shift to spread a mattress and lie on that instead of the hard floor, my eyes all the while fastened on my strange guest, who, remarking again that I would have 'only pains and calentura', chuckled as he chanted a wild song:

> High are the waves, fierce, gleaming,
> High is the tempest roar!
> High the sea-bird screaming!
> High the Azore!

I suppose I was now on the mend, for I was peevish, and complained: 'I detest your jingle. Your Azore should be at

roost, and would have been were it a respectable bird!' I
begged he would tie a rope-yarn on the rest of the song, if
there was any more of it. I was still in agony. Great seas were
boarding the *Spray*, but in my fevered brain I thought they were
boats falling on deck, that careless draymen were throwing from
wagons on the pier to which I imagined the *Spray* was now
moored, and without fenders to breast her off. 'You'll smash
your boats!' I called out again and again, as the seas crashed
on the cabin over my head. 'You'll smash your boats, but you
can't hurt the *Spray*. She is strong!' I cried.

I found, when my pains and calentura had gone, that the
deck, now as white as a shark's tooth from seas washing over it,
had been swept of everything movable. To my astonishment, I
saw now at broad day that the *Spray* was still heading as I had
left her, and was going like a race-horse. Columbus himself
could not have held her more exactly on her course. The sloop
had made ninety miles in the night through a rough sea. I felt
grateful to the old pilot, but I marvelled some that he had not
taken in the jib. The gale was moderating, and by noon the sun
was shining. A meridian altitude and the distance on the patent
log, which I always kept towing, told me that she had made a
true course throughout the twenty-four hours. I was getting
much better now, but was very weak, and did not turn out reefs
that day or the night following, although the wind fell light; but
I just put my wet clothes out in the sun when it was shining, and,
lying down there myself, fell asleep. Then who should visit me
again but my old friend of the night before, this time, of course,
in a dream. 'You did well last night to take my advice,' said he,
'and if you would, I should like to be with you often on the
voyage, for the love of adventure alone.' Finishing what he had
to say, he again doffed his cap and disappeared as mysteriously
as he came, returning, I suppose, to the phantom *Pinta*. I
awoke much refreshed, and with the feeling that I had been in
the presence of a friend and a seaman of vast experience. I
gathered up my clothes, which by this time were dry, then, by
inspiration, I threw overboard all the plums in the vessel.

7
Storm in the South Atlantic

In 1923 Conor O'Brien left Dublin in his yacht *Saoirse* of twenty tons
Thames measurement, and sailed round the world by way of Cape
Horn. He tells us that he started on the voyage because he had been
invited to join a mountaineering party in New Zealand; and, having
a nearly new yacht, he thought that he could both try her out and
reach New Zealand by sailing there in *Saoirse*. In the event, he
arrived too late to climb. He met bad weather before the Cape of
Good Hope and I have taken his description of this part of his passage as an example of South Atlantic weather. *Saoirse* had heavy gear
and could set a large square sail; however, O'Brien usually had a
strong crew whom he engaged at the ports where he called.

A. R.

from 'ACROSS THREE OCEANS'
by CONOR O'BRIEN

ON the following day we lost the Trades in lat. 22° S.,
and very soon our surroundings began to change. The
increase in bird life was most conspicuous. The avifauna of the
Tropics is very poor; a few Bosun birds and a solitary gannet
who had probably strayed north on a fishing excursion from
Trinidad were all I noticed; but now albatrosses were seen by
twos and threes, and Cape pigeons by scores. We had seen the
last of the flying-fish, but bonitos and dolphins were still with us,
feeding, I suppose, on something else which I did not see and
could not catch; their beat was shared by an occasional whale
and innumerable porpoises; all in increasing numbers as we
drew to the southward.

We went slowly enough at first, as the winds were still very
light, though what there was came from the north-east, and the
sea stayed always smooth; we sent up the foreyard and bent and

set the sail, and went a little faster; on the third day we were
making over 6 knots with a moderate breeze, still from the
same quarter; on the fifth, in lat. 30° S. and long. 20° W., the
Devil tempted me and I fell. He can put a very specious appearance on his temptations; the wind had gone more northerly,
there was a considerable swell, and admittedly we were running
to the eastward of our proper course; and on that 16th day of
September there was no harm done. Only on the following day I
should have made more southing, even at the cost of taking in all
the fore-and-aft canvas. The tempter hinted that as we were so
near Table Bay (1,750 miles, to be precise), 150 miles a day in
that direction were better than 100 in any other; that what we
wanted was gentle northerly breezes and not westerly gales;
that we had lit the galley stove for the first time that day on the
verge of the Tropics, and what should we feel like any farther
south? and, lastly, that if we went too near to Tristan da Cunha
the inhabitants would come out and massacre us for omitting to
bring their mail. So next day we found that we had so much
more easting than southing that we had to hypothecate a strong
current to account for our position; when the Devil puts his
hand to a job he does it thoroughly. And even the day after
we were sailing so fast and easily that I let her go on the same
course. When at last I woke up the fact that I had not yet
reached the 33rd parallel and was actually getting set to the
northward of east, it was too late to remedy my mistake, as it
was blowing up for a southerly gale, with a rising sea. By the
evening of the 20th of September we had sent down the foreyard and put two reefs in the mainsail, and were reaching along
fairly comfortably, though slowly enough.

The earlier part of this day was splendid sailing. There was a
big swell coming down from the North, the relic of the good
breezes of the past two days, and a big and increasing sea running up from the South, but not so much wind behind it as to
make it dangerously steep or breaking. With these opposing sets
of waves there was no oscillatory movement of the water; the
ship merely rose and fell, 20 feet at a time, going along as

steady as a racing yacht in the Solent, with the foam just lapping the lee covering-board. There is no pleasanter sensation than sailing along a really big beam sea, but it is one in which one can seldom indulge, and only for a short time. Soon we had to reduce sail, and haul our wind so as to take the seas on the bow; for by now the old swell was nearly extinguished and the new one was running into steep breaking crests. Next morning it was blowing a gale, right ahead on our course; I had taken in jib and mizzen, and we were head-reaching under staysail and mainsail, both close-reefed. Neither the wind nor the sea was worse than what one expects off our own coasts; the nuisance was that it was a foul wind. A foul wind in the open ocean is, however, generally a remediable nuisance, so I gave up sailing very slowly in a considerably wrong direction, and went off at large to look for a fair wind.

Before embarking on the next paragraph, in which I shall have to use some words of strange meaning, I must explain that there is no term in the English language that denotes unequivocally a system of winds rotating round an area of low pressure. The depression beloved of the weather forecaster does not quite fill the bill; besides, it is ambiguous to say 'The depression has passed away from us.' So I fall back, as the meteorologist does, on 'cyclone' for the whole system, and, for brevity, 'storm centre' for the middle of it, without implying any particular strength of wind. For this purpose I use the word 'gale' suitably qualified. Thus 'half a gale' means a wind blowing up to 30 miles an hour; a 'gale' to 40; a 'fresh gale', 'whole gale', 'strong gale' may approach 50, but is generally a simple 'gale' complicated by cold rain or a catastrophe in the galley.

It was obvious that the storm centre was somewhere to the northward; it was probable that it was moving to the eastward; but I had no indication of how far away it lay, and the barometer was very sluggish and unhelpful. But if I kept the ship running before the wind I should eventually get a westerly breeze, and I supposed the barometer would tell me if I were getting imprudently near the centre. It did not; this depression was, like most extra-tropical cyclones, a saucer-shaped affair; and as we proceeded to the northward the wind actually took off a little. In the evening the dull and rather thick weather cleared, and I saw a great commotion in the upper clouds and deduced that we were pretty near the centre. The whole affair must have been quite stationary, or even moving in the wrong direction. While the wind was still tolerably steady we hove-to and had dinner, and then waited to see what would happen.

It fell a calm, then the waves, no longer controlled by any order, ran in every direction, climbed upon each other's backs, and hurled misshapen lumps of black water into the air; at times it felt as if they were hurling the ship into the air as well. We had a good view of all this, for right above our heads the moon hung in a small clear space of sky. Round her black clouds, throwing out here and there streamers of brilliant white, whirled in a mad dance, at one moment retreating almost to the horizon which was lurid with lightning, at another rushing towards the charmed centre. At last one invaded it, and there burst into rain. It was overwhelming, stunning. I at the wheel could not see the compass a foot away. The mate, from the shelter of the chart-room, bawled out instructions to me, for I wanted to get steerage way on the ship as soon as possible. Though the wind had eased off gradually, it might come on again with a bang and knock the mast out of us. I do not know just when the wind did come; very little of it was enough to give me command of the ship, for that deluge had knocked the sea flat; and it came gradually, with intervals of calm and comparatively light showers. It was not the conventional behaviour of a cyclone, but since the dirt started the barometer had only dropped fifteen-hundredths of an inch.

Daybreak saw us running before a fresh westerly breeze, and I set the jib and turned one reef out of the mainsail. I should have turned out the crew and sent up the foreyard, but I felt we all needed a bit of rest, and sending up that foreyard was likely to be an all-hands and all-the-morning job. We had not shipped any heavy seas, so it must have been the rain that washed all the gear into an inextricable tangle. By the time that things had dried out and been to some extent cleared up, there was far too much wind and sea to set the sail; for we had now got on the right side of the weather and into a smart westerly gale.

This was my first experience of easting weather. I had sworn, in these circumstances, to take in my mainsail and run under square canvas only. For to a small vessel sea is usually a more serious matter than wind, and if it is really big she has to steer dead before it, and chance a gybe. Being without a main boom we could take that chance, but naturally preferred a rig which eliminated all risk. Today, however, since I could not set any square canvas, I wanted to avoid reducing the fore-and-afters, nuisance as they were in a heavy sea; for I was not wasting a fair wind. I let the ship run on under two headsails and single-reefed mainsail. I wanted to bring the wind as far out on the starboard quarter as I dared, to make up some of that southing I had lost the day before, and that kept all the sails full, so they were not so much of a nuisance as I had expected. But looking back on it, the sea was not so bad as I thought it was; the ship, however, took a lot of steering, and I took P's watch for him. Part of the difficulty of steering was due to the cramped positions of the helmsman, but part was due to the benefactions of a friend in Dublin, who had insisted that the wheel chains were not strong enough, and had rove off a stouter chain. Meanwhile the unlucky P. had kicked the original chain overboard. The new chain, naturally enough, jammed in the sheaves when a strain came on it, and, such is the perversity of inanimate things, jammed worst at such moments as I, fearing the ship would broach to, wanted to put the helm hard up. With my heart in my mouth I had to let the wheel run back a spoke or two and have another try at it. I dared not force it for fear of breaking something. But all that evening we ran unscathed.

Dark mountain ranges, their lower slopes netted with the tracery of spindrift, reared snowy crests against the sky; from them jutted sharp-cut spurs, which, growing with incredible rapidity, exploded against the bulwarks and discharged a salvo of spray into the belly of the sails. The ship, her jib-boom sweeping the water already mottled by the fountain thrown from her cut-water, rose on the steep face of the advancing swell and drowned the lighter patter of the drops torn from its summits with the shrieking in the rigging of the now unimpeded wind; then with a great crash the sea spread out under her in a boiling sheet of white. On this, or, as it seemed, suspended over this, and uncontrolled by contact with the firm support of the waters, she flew for a long space, then gradually settled down on the streaked back of the roller; the singing of the wind in the stays dropped to a low murmur, and in an uncanny quiet she shook herself and prepared for the next onset. So hour after hour the rhythm goes on, a hissing, rattling, shrieking crash; and then the bubbling of foam; a hypnotic influence. H. said that I was asleep when he relieved me at the wheel, but I do not believe that was literally true.

I have very frequently been asked whether my sails, which on a day like this would be little more than twenty feet above the water level, were not becalmed between the seas. What I have described in the last paragraph certainly gave the illusion of being becalmed; but it was, I imagine, only an illusion, due to the facts that on the back of a wave one is undoubtedly moving slowly because one is sailing uphill, and that one is in smooth after a very noisy and agitated few moments. It is physically impossible that the wind should be lighter on this, the windward, slope of the sea, where one might expect a calm would be close under the crest; but here, it seemed to me, probably because it was puffy, that it blew the hardest. In connexion with this I noticed one fact that surprised me very much, that there was always a smooth depression in the crest immediately astern of us (I was not using oil, it was just Providence) and one might expect a strong wind to blow over that as it would over a mountain pass. But the fact that I lost no sails running in heavy weather proves that the puffs cannot have been very heavy or the soft spots very light.

We had now managed to get back to lat. 32° S., but in long. 4° W. the wind left us again. Properly we should have been somewhere about the 40th parallel, though I learned afterwards that conditions had been abominable down there; anywhere between was full of easterly winds. But it was too late to try to get South now; I had to make the best of the latitude I was in.

As a matter of fact I made a poor use of one fine day, on which I logged a very light westerly breeze and a confused swell. I had all the fore-and-afters in and drifted along under foresail only, while all the clothes and bedding, books and potatoes were hauled up on deck. We seem to have indulged in this amusement far more often in the early stages of the voyage than later, much to the detriment of our sailing record, whether because the decks were more leaky then, or because we were still struggling with the awful legacy of that first three weeks' culture of mildews I do not know. Even crossing the Indian Ocean, with a crew none of whom was conspicuous for the domestic virtues, though the ship was indescribably filthy it was with inoffensive filth; and of course after I had shipped a Tongan crew everything was beautifully clean. At this stage, however, there was mighty little result to show for our labours, except an unnecessarily slow passage. To be quite just, though, I ought not to forget that the wind was right aft, very light, and with a confused swell; as happened at times also in the Indian and Pacific Oceans, where, too, the fore-and-afters had occasionally to be taken in; but then I had half as much square canvas again, the foretopsails, and those so light that an air would fill them. As things were, I had soon to put the potatoes back in their locker, the books in the shelves, the bedding on the bunks, and the clothes on my back (for it was getting cold and windy) and turn my attention to dealing with the weather according to the rules of Storm Sailing; which meant a hopeful attempt to keep the ship in that quadrant of the approaching cyclone which provided a fair wind for Table Bay.

Now while it is generally true of the South Atlantic that the winds are easterly in the Tropics, northerly off Brazil, westerly to the southward, and southerly up the African Coast, there are modifications in detail. Nowhere can one be sure of a westerly wind; even in 55° S. I have met a North-Easter. One does not run down one's easting with the wind all the time right aft, but with it anywhere from north to south-west; the rapidity of its changes depending on the distance of the ship from the centres of the depressions, and their relative speeds. I had located one depression to the southward, unpleasantly near, and moving irregularly; which meant shifty winds and a consequently cross-sea. These things generally travel in company; all I could do was to stay where I was and hope that the next one that came along would move with the same speed and in the same direction as myself. If it travelled, as it properly should, towards Cape Agulhas at 7 knots, we should be in Table Bay in three days. But wherever it went I was determined to keep to the northward of it, and chance head winds and a foul current later on. It started nicely on the 28th of September with a moderate north-west gale, for which we were ready, stripped to jib and foresail.

This day I noticed quite unprecedented swarms of birds: albatrosses, mollymawks, giant petrel, Cape pigeons, Mother Carey's chickens, and most profusely, prions. It is said that a profusion of these in the middle latitudes indicates severe weather farther south. If so, I was well off where I was; the weather just suited me, and I did not want it any more severe. This time we were running dead before the sea and could admire its majesty without being terrified by its steepness; all of us except the helmsman, that is; for there is a law of the Medes and Persians that he may not look behind him. So long as he kept the ship straight she steered so easily that broaching-to was unthinkable; but if he saw those blue cliffs rolling up astern he might lose his head, and our lives. By way of encouragement to P., I slept under his hand, so to say, in the chart-room during his watch; not perhaps the wisest thing in the world, for if we got pooped, and one must always remember the possibility of the most unlikely accident, the chart-room would certainly be filled up and I should probably be drowned. For there was a big sea that night. I have seen bigger off Eagle Island in Mayo, and off Cape Leeuwin in Australia, but only one of each; there were eight hours of these. There is, however, something about my yacht, probably her very clean run, that tames the ugliest-looking following sea, and we did not take a drop of water on deck, except of course in the way of that unlucky galley sky-light.

8

How Cape Horn Got It's Name

Many of the stories in this book are about the seas round Cape Horn. As Everest is to the mountaineer, so this cape is to the sailor, and this promontory holds a very romantic place in the imagination of man, whether he goes to sea or not. It is not necessary to go round Cape Horn to pass between the Pacific and the Atlantic, as there are two passages to the north which cut the corner. The best-known is Magellan Strait, which was the normal route taken by ships, but even if it gives a respite from the Southern Ocean, it is difficult and dangerous.

The voyage here described and the discovery of the Horn came about because the use of the Magellan Strait had been forbidden to all ships except those of the Dutch East India Company, just as in our own time the closing of the Suez Canal has forced ships to take the longer Cape route.

A. R.

from 'CALLANDER'S VOYAGES'
by JOHN CALLANDER, d. 1789

THE States General of the United Provinces having granted to the East-India Company an exclusive charter, prohibiting thereby all their subjects, except the said Company, from carrying on any trade to the eastward beyond the Cape of Good Hope, or westward through the Straits of Magellan, in any countries either known or unknown, under very high penalties; this prohibition gave very great distaste to many rich merchants, who were desirous of fitting out ships and making discoveries at their own costs, and could not help thinking it a little hard, that the Government should thus, against the laws of nature, bar those passages which Providence had left free. Amongst the number of these merchants, there was one of Amsterdam, who

then resided at Egmont, very rich, well-acquainted with business, and who had an earnest desire to employ a part of that
wealth, which he had acquired by trade, in acquiring fame as a
discoverer. With this view he applied himself to William Cornelison Scouten, of Horn, a man in easy circumstances, and who
was deservedly famous for his great skill in maritime affairs,
and for his perfect knowledge in the trade to the Indies, having
been thrice there himself, in the different characters of master,
pilot, and supercargo, or, as the phrase in those days was, of
Merchant. The great question proposed by Mr le Maire to this
intelligent man was, Whether he did not think it possible to
find another passage into the South Seas than by the Straits
of Magellan; and whether, if this was possible, it was not highly
likely, that the countries to the south of that passage might
afford as rich commodities as either the East or West Indies?
Mr Scouten answered, That there was great reason to believe
such a passage might be found, and still stronger reason to
confirm what he conjectured as to the riches of these southern
countries.

After many conversations upon this subject, they came at last
to a resolution of attempting such a discovery, from a full
persuasion, that the States General could not intend, by their
exclusive charter to the East-India Company, to preclude their
subjects from discovering countries on the south by a new route,
distinct from either of those mentioned in that charter. In
consequence of this agreement, it was stipulated, that le Maire
and his friends should advance one moiety towards the necessary
expence of the voyage, and Schouten and his friends the other.
In pursuance of this scheme, Isaac le Maire advanced his part of
the money; and Cornelison Scouten, with the assistance of the
following persons, viz. Peter Clementson, Burgomaster of
Horn; John Janson Molenwert, one of the Schepen or Alder-
men of the same place; John Clementson Keis, Senator of the
said town; and Cornelius Segetson, a merchant of Horn, laid
down the rest. It is certain that so many people of substance
would never have embarked in such a project if they had so

much as suspected that the East-India Company had a right to confiscate their vessels and effects whenever they had it in their power.

As soon therefore as these matters were adjusted, which was in the spring of the year 1615, the Company, engaged in this undertaking, began to apply themselves to the carrying it into execution, proposing to equip for that purpose a larger and a less vessel, to sail from Horn at the proper season of the year. And that all parties might be thoroughly satisfied, it was determined, that William Cornelison Schouten, on account of his age and experience, should have the command of the larger ship, with the sole direction of the voyage, and that James le Maire, the eldest son of Isaac le Maire, should be the first supercargo. The Company were so eager in the prosecution of their design, and so attentive to whatever might be necessary to promote it, that in the space of two months all things were ready, and a sufficient number of men engaged for navigating both ships. But, as secrecy was absolutely necessary, the seamen were articled in general terms to go wherever their masters and supercargoes should require; and, in consideration of so unusual a condition, their wages were advanced considerably; which was a circumstance of such consequence, and there were in those days so many adventurous spirits, that they did not find it at all difficult to make up their intended complement; which gave them an opportunity of choosing none but experienced mariners, on whose skill and fidelity they could depend; a circumstance of the utmost consequence in a voyage of this nature, where the tempers of men were sure to be thoroughly tried.

These extraordinary preparations, but, above all, the mighty secrecy that was observed, caused a great noise, not only at Amsterdam, but all over Holland, where several people reasoned on the intention of this voyage, according to the several degrees of their capacity and experience, some fancying they were bound to one place, some to another; but the common people thought they hit upon the proper title, in calling them the

Gold-finders; whereas the merchants, who were better versed in such matters, called them, with greater propriety, the South Company, and indeed that was their true designation; for the real design of Isaac le Maire, was to discover those southern regions, to which few people had hitherto travelled even in imagination, and which, by an unaccountable indolence, remain, in a great measure, undiscovered to this day.

In the beginning of the month of May 1615, the South Company drew their men together; and, on the 16th of that month, they were mustered before the magistrates of Horn, took their leave of their friends and relations, and prepared to embark on board their ships.

The biggest of these vessels was called *The Unity*, of the burden of 360 tons, carrying nineteen pieces of cannon, and twelve swivels. She had on board likewise a pinnace to sail, and another to row, a launch for landing of men, and a small boat, with all other necessaries whatever for so long a voyage; and of this vessel William Cornelison Schouten was master and pilot, and Jaques le Maire supercargo. The lesser ship was called *The Horn*, of the burden of 110 tons, carrying eight cannon and four swivels, John Cornelison Schouten master, and Aris Clawson supercargo. The crew of the former consisted of sixty-five men, and the latter of twenty-two only. The *Unity* sailed May 25th, for the Texel, where the *Horn* likewise arrived June 3rd following, that being judged the properest season of the year for them to proceed on their voyage. On June 14th they sailed out of the Texel, and passing, in sight of Dunkirk, between Dover and Calais, anchored on the 17th in the Downs, when William Cornelison Schouten went on shore at Dover, in order to get fresh water, and to hire an English gunner; which accordingly he did, and that day sent him on board. They sailed again in the evening, and met with several large Dutch ships laden with salt. In the night between the 21st and 22nd, they were grievously ruffled by a storm; which obliged them to put into the Isle of Wight for shelter, where Captain Schouten endeavoured, if possible, to have hired a carpenter, but without success, which obliged them to sail on the 25th for Plymouth, where he arrived on the 27th, and there hired a carpenter of Maydenblick.

On July 28th, they sailed from Plymouth with a north-north-east wind, and very fair weather. On the 29th, Captain Schouten made a signal for all officers to come on board; when it was resolved in a council, to settle the rate of their sea-allowance in such a manner, as that the men might have no reason to complain, and their officers be in no apprehensions of their wanting provisions during the course of so long a voyage. The rate they fixed in the following proportions; viz, a can of beer a man every day, four pounds of biscuit, half a pound of butter, and as much sweet suet, for the week, together with five large Dutch cheeses, that were to serve them the whole voyage. This was exclusive of flesh or fish: And we may, from hence, form some notion of the frugality the Dutch victualled with in those days, and from which they have deviated very little ever since. They likewise made the necessary orders for the due regulation of the voyage, directing, that, in case of landing men, one of the masters should always command; that, in ports where they went to trade, the supercargo should go on shore, and have the sole direction of the commerce: That, on board, every officer should be strict in the execution of his duty; but without putting unnecessary hardships on the men, or interfering with other officers in their commands: That none of the officers should hold any conversation with the seamen, in relation to the design of the voyage, which being solely in the breast of the first captain and supercargo, conjectures must be fruitless, and might be dangerous; that any embezzlement of provisions, stores, or merchandise, should be severely punished; and, in case of their being reduced to short allowance, then offences of this nature to be punished with death; that the two supercargoes should keep clear and distinct journals of all proceedings, for the use of the Company, that it might plainly appear, how far every man had done his duty, and to what degree the end of the voyage had been answered. All these rules were very exactly observed, and particularly the last; so that, from these journals, kept by the supercargoes, this account has been taken. On July 8th, being in the latitude of 39° 25′, their carpenter's mate died. On the 9th and 10th, with a north-north-east wind, and a stiff gale, they stood on their course, without putting in to Porto Santo, or Madeira, of which they had sight on the 11th. The reason was, that having, as they conceived, victuals sufficient for the voyage, they determined not to lose time, by going needlessly on shore, especially since hitherto their men were vigorous, and in good health; which resolution was founded on an observation made by Captain Schouten, that many voyages had been lost, by lingering in port without any urgent cause, when the winds and seasons were fair, and their course might have been prosecuted without delay.

On July 13th, they sailed between the island Teneriff and the Grand Canary, with a stiff north-north-east wind, and a swift current. About the 15th, the same wind and current following them still, they passed the Tropic of Cancer. The 20th, in the morning, they fell in with the north side of Cape Verd. At sun-rising the Cape lay west by south from them; so that the north-north-east wind would not suffer them to get beyond it; but kept them there at anchor all that night. The 25th, the Moorish Alcaid came on board them, with whom they agreed at the price of eight states of iron for a supply of fresh water. They left the Cape August 1st, and the 21st of the same month they saw the high land of Sierra Leona, and also the island of Madrabomba, which lies on the south point of the high land of Sierra Leona, and north from the shallows of St Anne's island. This land of Sierra Leona is the highest of all that lies between Cap Verd and the coast of Guinea, so that the point is very easy to be known. Here they would have landed, running up to the point over the Baixos, or shallows, of St Anne's at ten, nine, eight, seven, and five fathoms water, it being still deeper to the north, but shallower to the east; so that in the evening, they anchored with a high-water at four fathoms and a half in soft ground, and at night at three fathoms and a half.

The 22nd, William Schouten, in the *Horn*, led the way off the shallows, steering north-north-east, with a north-west wind, by which course they were intirely disengaged from the Baixos, and got into thirteen fathoms water. From hence they went to the islands of Madrabomba, which are very high, and lie all three on a row south-west and north-east, half a league from Sierra Leona, to the seaward. Here they had shallow water at four and five fathoms, and soft muddy ground. They anchored a league from the island, which appeared to be very full of bogs and marshes, and all over waste like a wilderness, scarce fit to entertain any other inhabitants than wild beasts, and indeed not seeming to have any other. Going ashore on the 23rd, they found a river there, the mouth of which was so stopped up with sands, and cliffs of rocks, that no ship could get into it; yet, within, the water was sufficiently deep, and the breadth such too, as to give a ship free scope to turn and wind herself about, as she should have occasion. Here they saw tortoises, crocodiles, monkeys, wild oxen, and a sort of birds, which made a noise, barking like dogs. They met with no fruit but lemons, some few trees of which they found after a tedious search. The 29th, about noon, they got above the islands of Madrabomba westward, along to the north part of the high land, till they had twelve and fifteen fathoms water, and, in the evening, got about the point.

On the 30th, being assisted both by wind and current, they arrived before the village that looks upon the road of Sierra Leona, where they anchored at eight fathoms water, a little from the shore, in a very sandy bottom. The village consisted of about eight or nine poor houses covered with straw: The Moors, that dwelt there, were willing to come aboard, only demanding pledges to be left ashore, to secure their safe return; because a French ship that came thither before, had perfidiously carried off two of them: So Aris Clawson the merchant went ashore, and stayed there amongst them, driving a small trade with them for lemons and bananas, which they exchanged for glass-beads; and in the meantime they came on board, bringing an interpreter with them, who spoke all sorts of languages. Here they had good opportunity of furnishing themselves with fresh water, which pouring down in large quantities from a very high hill, they had nothing to do but to place their barrels under the fall of the water to receive it. There were also vast woods of lemon-trees here, which made lemons so cheap to them, that for a few beads and knives, they might have had 10,000. September 1st, they drove away before the stream, and anchored that evening at the mouth of the sea, before a small river. Here they took an antelope in the woods, with lemons and palmitoes; and had good success in their fishing. The 3rd the master brought in a great shoal of fish, that were of the shape of a shoemaker's knife, and as many lemons as came to 150 for every man's share.

The 4th, they sailed from Sierra Leona, early in the morning. October 5th, they made 4 degrees 27 minutes south latitude; and the same day at noon, they were strangely surprised with a very violent stroke given to one of their ships in the lower part of it. No adversary appeared, no rock was in the way to be encountered with; but, while this amused them, the sea all about them began to change its colour, and looked as if some great fountain of blood had been opened into it; this sudden alteration of the water being no less surprising to them, than the striking of the ship; but the cause both of the one and the other they were equally ignorant of, till they came to Port Desire, and there set the ship upon the strand to make her clean; for then they found a large horn, both in form and magnitude resembling an elephant's tooth, sticking fast in the bottom of the ship. A very firm and solid body it was, and seemed to be equally so all over, there being nothing of a cavity, or a light and spungy matter in the midst of it, but all over as dense and compact a substance, as that in the exterior parts. It had pierced through three very stout planks of the ship, and razed one of the ribs of her; so that it stuck at least half a foot deep in the planks; and there was about as much that appeared without the great hole up to the place where it was broken off. And now the riddle was completely solved, this horn being the spoil of some sea-monster, that had thus rudely assaulted the ship with that piercing weapon; and, after the thrust, not being able to draw it out again, had there broke it, which was attended with such a plentiful effusion of blood, as had discoloured the sea to that degree.

Having now sailed so far, that none in the ships, but the master, knew where they were, or whither they intended, upon the 25th, they discovered their designs to the rest of the company, of going to find out a new southern passage into the great Pacific Sea. This they had kept very close to themselves before, but now thought it time to reveal the scheme, there being no danger of defeating it; and the company seemed to be very well pleased with it, hoping to light on some golden country or other, to make them amends for all their trouble and danger. The 26th, they made 6 degrees 25 minutes south latitude, sailing mostly southwards, till they had made 10 degrees 30 minutes. November 1st, they had the sun north of them at noon. The 3rd, in the afternoon, they had sight of Martin Vad's island, called Ascension, under 20 degrees; and here they observed the compass to vary to the north-east 12 degrees. The 21st, they came under 38 degrees 25 minutes, and had a deep water, whose bottom they could reach with their lead. Here the variation of the compass was 17 degrees to the north-east. December 6th they had a prospect of land, not very high, but flat and white; and, quickly after, fell in with the north side of Port Desire, and, that night, anchored within one league and an half from the shore, in ten fathoms water, with an ebb that ran southwards as strongly as the sea runs between Flushing Heads.

The 7th, keeping a south course, at noon they came before the haven of Port Desire, which lies under 40 degrees 40 minutes. At the entry of it they had very high water; neither did any of those cliffs appear, which Van Noort had described, and which he left northwards in sailing into the haven. If there were any, they were all under water; but the cliffs lay open and visible enough towards the south point, which therefore, might be those which Noort intended. Upon this they went on, sailing so far southward, as to miss the right channel. They came into a crooked bay, where, at high-water, they had but four fathoms and a half, and at low water but fourteen feet; by which means the *Unity* lay with her stern fast aground, and, if a brisk gale from the north-east had blown, she must infallibly have been lost; but, the wind blowing west from the land, she recovered again. Here they found plenty of eggs, amongst the cliffs: and the bay afforded them muscles, and smelts of sixteen inches in length, and therefore they called it Smelt-Bay. Their shallop went to the Penguin islands, and came back with 150 penguins, and two sea-lions.

The 8th, before noon, they sailed out of the Smelt-Bay, and anchored just before Port Desire. The shallop was employed before hand to sound the depth of the channel; which proving to be twelve or thirteen fathoms, they boldly entered, having a north-east wind to carry them along: But after a little more than a league's sailing, the wind began to veer about, and they anchored at twenty fathoms; but the bottom they were upon being only slippery stones, and the wind now blowing hard at north-west, their anchors could not preserve them from driving with that rough wind upon the southern shore; so both these ships were likely to be wrecked together. The *Unity* lay with her side upon the cliffs; but still kept the water, and, by the fall of the sea, was gradually sliding down lower and lower into it; but the *Horn* stuck so that her keel was above a fathom out of water, and a man might have walked dry under it at low water. She was, for some time, obliged to the north-west wind, that, by blowing hard upon her side, kept her from falling over; but, that support being done, with the wind that gave it, she sunk down upon that side at least three feet lower than the keel: Upon which sight they gave her over for lost; and yet the succeeding flood, which came on with still weather, set her upright again; and both she and her companion got clear of that danger. The 9th, they went farther into the river, and came to King's island, which they found full of black sea-mews, and almost covered over with their eggs. A man, without straining to reach, might have taken between fifty and sixty nests with his hand, each of which have three or four eggs a-piece; so that they were quickly furnished with some thousands of them. The 11th the boat went in search of good water lower down the river, on the south side; but found it all of a brackish unpleasant taste. They saw ostriches here, and a sort of beasts like harts, with wonderful long necks, and extremely wild. Upon the high hills they found great heaps of stones, under which some monstrous carcases had been buried. There were bones of ten and eleven feet long. In all probability they were (if of rational creatures) some bones of the giants of that country. No water was to be found here for several days together; so that, though they had plenty of good fish and fowl, they could meet with no drink to wash it down.

On the 17th, they laid the *Unity* dry upon King's Island in order to clean her, which they performed very successfully. On the 18th, they likewise hauled the *Horn* on shore, for the same purpose, and placed her about 200 yards from the other ship. On the 19th, a very dreadful accident happened; for, while they were busy cleaning both ships, in order to which it was necessary to light a fire of dry reeds under the *Horn*, it so fell out, that the flame got into the ship, and set it on fire; and, as they were fifty feet from the water-side, they were forced to stand still and see her burn, without being able to do anything towards extinguishing it. On the 20th, at high water, they launched the *Unity*, and the next day carried on board her all the wood, iron-work, anchors, and pieces of cannon, and whatever else they could save out of the *Horn*. On the 25th, the sailors found certain holes full of fresh water, which was white and thick, but well tasted, a great quantity of which they carried on board in small casks upon their shoulders. They met here with great numbers of sea-lions; the young ones they ate, and found them pretty good food. The sea-lion is a creature as big as a small horse; their heads resemble lions exactly; on their necks they have long manes of a tough strong hair; but this is to be understood of the he-lions only: For the she-lion is without hair and not above half as big as the male. They are a bold fierce animal, not to be destroyed but by musket-shot.

January 13th, they sailed out of Port Desire; but having a calm, they anchored before the haven, till the rising of the wind invited them to pursue their voyage. The 18th, being in 51 degrees, they saw the Sebaldin islands; which they observed to lie in that position and distance from the Strait, that De Weert had determined. The 20th, being then in 53 degrees, they observed the great current that runs south-west; and now they reckoned about twenty leagues to the southwards of the Magellanic Straits. The 23rd they had an uncertain shifting wind, and the water appeared white, as if they had been within the land. They held their course south by west, and the same day saw land, bearing west and west-south-west from them, and quickly after to the south. Then attempting, by an east-south-east course, to get beyond the land, the hard north wind, that blew then, constrained them to take in their topsails. The 24th, in the forenoon, they saw land a starboard, about a league's distance, stretching out east and south, with very high hills, all covered with ice; and then other land bearing east from it, high and rugged as the former. They guessed the lands they had in these two prospects lay about eight leagues asunder, and that there might be a good passage between them, because of a pretty brisk current, that ran southwards along by them. About noon, they made 54 degrees 46 minutes, and then began to make towards the aforementioned opening; but the succeeding calm prevented it. Here they saw an incredible number of penguins, and such huge shoals of whales, that they were forced to proceed with great caution for fear they should run their ship upon them.

The 25th, in the forenoon, they got up close by the east land, which, upon the north side, reaches east-south-east as far as the eye can follow it. This they called States Land; and to that which lay west, they gave the name of Maurice Land. They observed there were good roads and sandy bays, good store of fish, penguins, and porpoises, and some sorts of fowls; but the land adjacent seemed quite bare of trees and woods. They had a north wind at their entrance into this passage, and directed their course south-south-west; so that, going pretty briskly on, at noon they made 55 degrees 36 minutes, and then held a south-west course, having a good stiff gale to blow them forwards. The land upon the south side of the passage, at the west end of Maurice Land, appeared to run west-south-west, and south-west as far as they could see it, and all very craggy uneven ground.

In the evening, having a south-west wind, they steered southwards, meeting with mighty waves, that came rolling along before the wind, and the depth of the water to the lee-ward from them, which appeared by some very evident signs, gave them a full assurance, that the great South Sea was now before them, into which they had almost made their way by a passage of their own discovering. The sea-mews thereabouts were larger than swans, and their wings, when extended to the full length, spread about the compass of a fathom. They would come and very tamely sit down upon the ship, and suffer themselves to be taken by hand, without any endeavours to fly away. The 26th, they made 57 degrees, and were there ruffled by a brisk storm out of the west and south-west. The water was also very high, and blue. They still held all this day their course to the southward, but changed it at night for a north-west one; in which quarter they discovered very high land. The 27th, they were under 56 degrees 51 minutes, the weather very cold, with hail and rain, the wind west and west by south. They went a southern course, and then crossed northwards with their main-sails. The 28th, they hoisted up their topsails, and had great billows out of the west, with a west and then a north-east wind, and therewith held their course south and then west and west by south, which brought them under 56 degrees 48 minutes.

The 29th they had a north-east wind, and held their course south-west which gave them the prospect of two islands, beset around with cliffs, and lying west-south-west from them; they got up to them at noon, but could not sail above them, and therefore held their course to the north. They gave them the name of Barnevelt's Islands, and found their latitude to be 57 degrees south. Taking a north-west course, from hence in the evening, they saw land again, lying north-west, and north-north-west from them; this was the high hilly land, covered with snow, that lay southward from the Magellanic Straits, ending in a sharp point, which they called Cape Horn, and lying in 57 degrees 48 minutes. They held their course now westward, in which course they found a strong current which ran that way too, yet had the wind in the north, and great billows rolling out of the west upon them. The 30th, the billows and the current still ran as before; and now they gathered a full assurance from hence, that the way was open into the South Sea: This day they made the latitude of 57 degrees, 34 minutes. The 31st, they sailed west, with the wind in the north, and made 58 degrees; but the wind turning to the west and west-south-west, they passed Cape Horn, losing all sight of land, and still meeting the billows rolling out of the west, which, together with the blueness of the water, made them quickly expect the main South Sea. February 1st, a storm blowing out of the south-west, they sailed with their mainsails north-west and west-north-west. The 2nd, with a westerly wind, they sailed to the southward, and made 57 degrees 58 minutes the variation being there 12 degrees northward. The 3rd, they made 59 degrees 25 minutes with a hard west wind, but saw no signs of any land to the south: and the next day 56 degrees 43 minutes, turning to and fro with very uncertain south-west winds, and finding 11 degrees of north-east variation. The 5th, by reason of a strong westerly current, and a hollow water, they could bear no sail, but were forced to drive with the wind.

The 12th they plainly discerned the Magellanic Straits, lying east of them; and therefore, now being secure of their happy new discovery, they rendered thanks to good Fortune in a cup of wine, which went three times round the company. And now this new-found passage had a name given it, which was that of Maire's Straits, though that honour (in justice) ought to have been done to William Schouten, by whose happy conduct the Straits were discovered.

9
Tierra Del Fuego

It was Joshua Slocum's writing which drew my attention to Darwin's description of Tierra del Fuego, which he read just after he has passed through the Milky Way. Darwin's voyage in the *Beagle* took him to the area of Cape Horn in 1833, and I have included it here because of his description of the people who inhabited the southern tip of South America. The *Beagle* had on board three Fuegians who had been taken to England on a previous voyage by Captain Fitz Roy to be educated and whom he now wished to re-patriate. Also an English missionary who planned to live and work among them.

A. R.

from *'A NATURALISTS VOYAGE*
ROUND THE WORLD'
by CHARLES DARWIN, M.A., F.R.S.

JANUARY 15th, 1883. The *Beagle* anchored in Goeree Roads. Captain Fitz Roy having resolved to settle the Fuegians according to their wishes, in Ponsonby Sound, four boats were equipped to carry them there through the Beagle Channel. This channel, which was discovered by Captain Fitz Roy during the last voyage, is a most remarkable feature in the geography of this, or indeed of any other country; it may be compared to the valley of Loch Ness in Scotland, with its chain of lakes and friths. It is about one hundred and twenty miles long, with an average breadth, not subject to any very great variation, of about two miles; and is throughout the greater part so perfectly straight, that the view, bounded on each side by a line of mountains, gradually becomes indistinct in the long distance. It crosses the southern part of Tierra del Fuego in an east and west line, and in the middle is joined at right angles on the south side by an irregular channel, which has been called

Ponsonby Sound. This is the residence of Jemmy Button's tribe and family.

January 19th. Three whale boats and the yawl, with a party of twenty-eight, started under the command of Captain Fitz Roy. In the afternoon we entered the eastern mouth of the channel, and shortly afterwards found a snug little cove concealed by some surrounding islets. Here we pitched our tents and lighted our fires. Nothing could look more comfortable than this scene. The glassy water of the little harbour, with the branches of the trees hanging over the rocky beach, the boats at anchor, the tents supported by the crossed oars, and the smoke curling up the wooden valley, formed a picture of quiet retirement. The next day (20th) we smoothly glided onwards in our little fleet, and came to a more inhabited district. Few if any of these natives could ever have seen a white man; certainly nothing could exceed their astonishment at the apparition of the four boats. Fires were lighted on every point (hence the name of Tierra del Fuego, or the land of fire), both to attract our attention and to spread far and wide the news. Some of the men ran for miles along the shore. I shall never forget how wild and savage one group appeared: suddenly four or five men came to the edge of an overhanging cliff; they were absolutely naked, and their long hair streamed about their faces: they held rugged staffs in their hands, and, springing from the ground, they waved their arms round their heads, and sent forth the most hideous yells.

At dinner-time we landed among a party of Fuegians. At first they were not inclined to be friendly; for until the Captain pulled in ahead of the other boats, they kept their slings in their hands. We soon, however, delighted them by trifling presents, such as tying red tape round their heads. They liked our biscuit; but one of the savages touched with his finger some of the meat preserved in tin cases which I was eating, and feeling it soft and cold, showed as much disgust at it as I should have done at putrid blubber. Jemmy was thoroughly ashamed of his countrymen, and declared his own tribe were quite different,

in which he was woefully mistaken. It was as easy to please as it
was difficult to satisfy these savages. Young and old, men and
children, never ceased repeating the word 'yammerschooner',
which means 'give me'. After pointing to almost every object,
one after the other, even to the buttons on our coats, and saying
their favourite word in as many intonations as possible, they
would then use it in a neuter sense, and vacantly repeat
'yammerschooner'. After yammerschoonering for any article
very eagerly, they would by a simple artifice point to their
young women or little children, as much as to say, 'If you will
not give it me, surely you will to such as these.'

At night we endeavoured in vain to find an uninhabited
cove; and at last were obliged to bivouac not far from a party
of natives. They were very inoffensive as long as they were
few in numbers, but in the morning (21st), being joined by
others, they showed symptoms of hostility, and we thought that
we should have come to a skirmish. A European labours under
great disadvantages when treating with savages like these, who
have not the least idea of the power of firearms. In the very act
of levelling his musket he appears to the savage far inferior to a
man armed with a bow and arrow, a spear, or even a sling.
Nor is it easy to teach them our superiority except by striking a
fatal blow. Like wild beasts, they do not appear to compare
numbers; for each individual, if attacked, instead of retiring,
will endeavour to dash your brains out with a stone, as certainly
as a tiger under similar circumstances would tear you. Captain
Fitz Roy on one occasion being very anxious, from good reasons,
to frighten away a small party, first flourished a cutlass near
them, at which they only laughed; he then twice fired his pistol
close to a native. The man both times looked astounded, and
carefully but quickly rubbed his head; he then stared awhile,
and gabbled to his companions, but he never seemed to think
of running away. We can hardly put ourselves in the position
of these savages, and understand their actions. In the case of
this Fuegian, the possibility of such a sound as the report
of a gun close to his ear could never have entered his mind. He

perhaps literally did not for a second know whether it was a sound or a blow, and therefore very naturally rubbed his head. In a similar manner, when a savage sees a mark struck by a bullet, it may be some time before he is able at all to understand how it is effected; for the fact of a body being invisible from its velocity would perhaps be to him an idea totally inconceivable. Moreover, the extreme force of a bullet that penetrates a hard substance without tearing it, may convince the savage that it has no force at all. Certainly I believe that many savages of the lowest grade, such as these of Tierra del Fuego, have seen objects struck, and even small animals killed by the musket, without being in the least aware how deadly an instrument it is.

January 22nd. After having passed an unmolested night, in what would appear to be neutral territory between Jemmy's tribe and the people whom we saw yesterday, we sailed pleasantly along. I do not know anything which shows more clearly the hostile state of the different tribes than these wide border or neutral tracts. Although Jemmy Button well knew the force of our party, he was, at first, unwilling to land amidst the hostile tribe nearest to his own. He often told how the savage Oens men, 'when the leaf red', crossed the mountains from the eastern coast of Tierra del Fuego, and made inroads on the natives of this part of the country. It was most curious to watch him when thus talking, and see his eyes gleaming, and his whole face assume a new and wild expression. As we proceeded along the Beagle Channel, the scenery assumed a peculiar and very magnificent character; but the effect was much lessened from the lowness of the point of view in a boat, and from looking along the valley, and thus losing all the beauty of a succession of ridges. The mountains were here about three thousand feet high, and terminated in sharp and jagged points. They rose in one unbroken sweep from the water's edge, and were covered to the height of fourteen or fifteen hundred feet by the dusky-coloured forest. It was most curious to observe, as far as the eye could range, how level and truly horizontal the line on the mountain side was at which trees ceased to grow; it

F

precisely resembled the high-water mark of drift-weed on a sea-beach.

At night we slept close to the junction of Ponsonby Sound with the Beagle Channel. A small family of Fuegians, who were living in the cove, were quiet and inoffensive, and soon joined our party round a blazing fire. We were well clothed, and though sitting close to the fire were far from too warm; yet these naked savages, though farther off, were observed, to our great surprise, to be streaming with perspiration at undergoing such a roasting. They seemed, however, very well pleased, and all joined in the chorus of the seamen's songs; but the manner in which they were invariably a little behindhand was quite ludicrous.

During the night the news had spread, and early in the morning (23rd) a fresh party arrived, belonging to the Teke-nika, or Jemmy's tribe. Several of them had run so fast that their noses were bleeding, and their mouths frothed from the rapidity with which they talked; and with their naked bodies all bedaubed with black, white and red, they looked like so many demoniacs who had been fighting. We then proceeded (accompanied by twelve canoes, each holding four or five people) down Ponsonby Sound to the spot where poor Jemmy expected to find his mother and relatives. He had already heard that his father was dead; but as he had had a 'dream in his head' to that effect, he did not seem to care much about it, and repeatedly comforted himself with the very natural reflection – 'Me no help it.' He was not able to learn any particulars regarding his father's death, as his relations would not speak about it.

Jemmy was now in a district well known to him, and guided the boats to a quiet pretty cove named Woollya, surrounded by islets, every one of which and every point had its proper native name. We found here a family of Jemmy's tribe, but not his relations; we made friends with them, and in the evening they sent a canoe to inform Jemmy's mother and brothers. The cove was bordered by some acres of good sloping land, not covered (as elsewhere) either by peat or by forest trees. Captain Fitz Roy originally intended, as before stated, to have taken York Minster and Fuegia to their own tribe on the west coast; but as they expressed a wish to remain here, and as the spot was singularly favourable, Captain Fitz Roy determined to settle here the whole party, including Matthews, the missionary, Five days were spent in building for them three large wigwams. in landing their goods, in digging two gardens, and sowing seed.

The next morning after our arrival (the 24th) the Fuegians began to pour in, and Jemmy's mother and brothers arrived. Jemmy recognized the stentorian voice of one of his brothers at a prodigious distance. The meeting was less interesting than that between a horse, turned out into a field, when he joins an old companion. There was no demonstration of affection; they simply stared for a short time at each other; and the mother immediately went to look after her canoe. We heard, however, through York, that the mother had been inconsolable for the loss of Jemmy, and had searched everywhere for him, thinking that he might have been left after having been taken in the boat. The women took much notice of, and were very kind to, Fuegio. We had already perceived that Jemmy had almost forgotten his own language. I should think there was scarcely another human being with so small a stock of language, for his English was very imperfect. It was laughable, but almost pitiable, to hear him speak to his wild brother in English, and then ask him in Spanish ('no sabe?') whether he did not understand him.

Everything went on peaceably during the three next days, whilst the gardens were digging and wigwams building. We estimated the number of natives at about one hundred and twenty. The women worked hard, whilst the men lounged about all day long, watching us. They asked for everything they saw and stole what they could. They were delighted at our dancing and singing, and were particularly interested at seeing us wash in a neighbouring brook; they did not pay much attention to anything else, not even to our boats. Of all the things which York saw, during his absence from his country, nothing seems more to have astonished him than an ostrich near Maldonado: breathless with astonishment he came running to Mr Bynoe, with whom he was out walking – 'O Mr Bynoe, oh! bird all same horse!' Much as our white skins surprised the natives, by Mr Low's account, a Negro-cook to a sealing vessel did so more effectually; and the poor fellow was so mobbed and shouted at that he would never go on shore again. Everything went on so quietly, that some of the officers and myself took long walks in the surrounding hills and woods. Suddenly, however, on the 27th, every woman and child disappeared. We were all uneasy at this, as neither York nor Jemmy could make out the cause. It was thought by some that they had been frightened by our cleaning and firing of our muskets on the previous evening; by others, that it was owing to offence taken by an old savage, who, when told to keep farther off, had coolly spit in the sentry's face, and had then, by gestures acted over a sleeping Fuegian, plainly showed, as it was said, that he should like to cut up and eat our man. Captain Fitz Roy, to avoid the chance of an encounter, which would have been fatal to so many of the Fuegians, thought it advisable for us to sleep at a cove a few miles distant. Matthews, with his usual quiet fortitude (remarkable in a man apparently possessing little energy of character), determined to stay with the Fuegians, who evinced no alarm for themselves; and so we left them to pass their first awful night.

On our return in the morning (28th) we were delighted to find all quiet, and the men employed in their canoes spearing fish.

Captain Fitz Roy determined to send the yawl and one whale boat back to the ship; and to proceed with the two other boats, one under his own command (in which he most kindly allowed me to accompany him), and one under Mr Hammond, to survey the western parts of the Beagle Channel, and afterwards to return and visit the settlement. The day, to our astonishment, was overpoweringly hot, so that our skins were scorched: with this beautiful weather, the view in the middle of the Beagle Channel was very remarkable. Looking towards either hand, no object intercepted the vanishing points of this long canal between the mountains. The circumstance of its being an arm of the sea was rendered very evident by several huge whales spouting in different directions. On one occasion I saw two of these monsters, probably male and female, slowly swimming one after the other, within less than a stone's-throw of the shore, over which the beech tree extended its branches.

We sailed on till it was dark and then pitched our tents in a quiet creek. The greatest luxury was to find for our beds a beach of pebbles, for they were dry and yielding to the body. Peaty soil is damp; rock is uneven and hard; sand gets into one's meat when cooked and eaten boat-fashion; but when lying in our blanket-bags, on a good bed of smooth pebbles, we passed most comfortable nights.

It was my watch till one o'clock. There is something very solemn in these scenes. At no time does the consciousness in what a remote corner of the world you are then standing come so strongly before the mind. Everything tends to this effect; the stillness of the night is interrupted only by the heavy breathing of the seamen beneath the tents, and sometimes by the cry of a night-bird. The occasional barking of a dog, heard in the distance, reminds one that it is the land of the savage.

January 29th. Early in the morning we arrived at the point where the Beagle Channel divides into two arms; and we entered the northern one. The scenery here becomes even grander than before. The lofty mountains on the north side compose the granitic axis, or backbone of the country, and boldly rise to a height of between three and four thousand feet, with one peak above six thousand feet. They are covered by a wide mantle of perpetual snow, and numerous cascades pour their waters, through the woods, into the narrow channel below. In many parts, magnificent glaciers extend from the mountain side to the water's edge. It is scarcely possible to imagine anything more beautiful than the beryl-like blue of these glaciers, and especially as contrasted with the dead white of the upper expanse of snow. The fragments which had fallen from the glacier into the water were floating away, and the channel with its icebergs presented, for the space of a mile, a miniature likeness of the Polar Sea. The boats being hauled on shore at our dinner-hour, we were admiring from the distance of half a mile a perpendicular cliff of ice, and were wishing that some more fragments would fall. At last, down came a mass with a roaring noise, and immediately we saw the smooth outline of a wave travelling towards us. The men ran down as quickly as they could to the boats; for the chance of their being dashed to pieces was evident. One of the seamen just caught hold of the bows as the curling breaker reached it; he was knocked over and over, but not hurt; and the boats, though thrice lifted on high and let fall again, received no damage. This was most fortunate for us, for we were a hundred miles distant from the ship, and we should have been left without provisions or firearms. I had previously observed that some large fragment of rock on the beach had been lately displaced; but until seeing this wave, I did not understand the cause. One side of the creek was formed by a spur of mica-slate; the head by a cliff of ice about forty feet high; and the other side by a promontory fifty feet high, built up of huge rounded fragments of granite and mica-slate, out of which old trees were growing. This promontory was evidently a moraine, heaped up at a period when the glacier had greater dimensions.

When we reached the western mouth of this northern branch of the Beagle Channel, we sailed amongst many unknown desolate islands, and the weather was wretchedly bad. We met with no natives. The coast was almost everywhere so steep that we had several times to pull many miles before we could find space enough to pitch our two tents; one night we slept on large round boulders, with putrefying seaweed between them; and when the tide rose, we had to get up and move our blanket-bags. The farthest point westward which we reached was

Stewart Island, a distance of about one hundred and fifty miles from our ship. We returned into the Beagle Channel by the southern arm, and thence proceeded, with no adventure, back to Ponsonby Sound.

February 6th. We arrived at Woollya. Matthews gave so bad an account of the conduct of the Fuegians that Captain Fitz Roy determined to take him back to the *Beagle*; and ultimately he was left at New Zealand, where his brother was a missionary. From the time of our leaving, a regular system of plunder commenced; fresh parties of the natives kept arriving; York and Jemmy lost many things, and Matthews almost everything which had not been concealed underground. Every article seemed to have been torn up and divided by the natives. Matthews described the watch he was obliged always to keep as most harassing; night and day he was surrounded by the natives, who tried to tire him out by making an incessant noise close to his head. One day an old man, whom Matthews asked to leave his wigwam, immediately returned with a large stone in his hand; another day a whole party came armed with stones and stakes, and some of the younger men and Jemmy's brother were crying; Matthews met them with presents. Another party showed by signs that they wished to strip him naked, and pluck all the hairs out of his face and body. I think we arrived just in time to save his life. Jemmy's relatives had been so vain and foolish, that they had shown to strangers their plunder, and their manner of obtaining it. It was quite melancholy leaving the three Fuegians with their savage countrymen; but it was a great comfort that they had no personal fears. York, being a powerful, resolute man, was pretty sure to get on well, together with his wife Fuegia. Poor Jemmy looked rather disconsolate, and would then, I have little doubt, have been glad to have returned with us. His own brother had stolen many things from him; and as he remarked, 'What fashion call that?' he abused his countrymen, 'all bad men, no sabe (know) nothing,' and, though I never heard him swear before, 'damned fools'. Our three Fuegians, though they had been only three years with civilized men, would, I am sure, have been glad to have retained their new habits; but this was obviously impossible. I fear it is more than doubtful whether their visit will have been of any use to them.

In the evening, with Matthews on board, we made sail back to the ship, not by the Beagle Channel, but by the southern coast. The boats were heavily laden and the sea rough, and we had a dangerous passage. By the evening of the 7th we were on board the *Beagle* after an absence of twenty days, during which time we had gone three hundred miles in the open boats. On the 11th, Captain Fitz Roy paid a visit by himself to the Fuegians, and found them going on well; and that they had lost very few more things.

On the last day of February in the succeeding year (1834) the *Beagle* anchored in a beautiful little cove at the eastern entrance of the Beagle Channel. Captain Fitz Roy determined on the bold and, as it proved, successful attempt to beat against the westerly winds by the same route which we had followed in the boats to the settlement at Woollya. We did not see many natives until we were near Ponsonby Sound, where we were followed by ten or twelve canoes. The natives did not at all understand the reason of our tacking, and, instead of meeting us each tack, vainly strove to follow us in our zigzag course. I was amused at finding what a difference the circumstance of being quite superior in force made, in the interest of beholding these savages. While in the boats I got to hate the very sound of their voices, so much trouble did they give us. The first and last word was 'yammerschooner'. When, entering some quiet little cove, we have looked round and thought to pass a quiet night, the odious word 'yammerschooner' has shrilly sounded from some gloomy nook, and then the little signal-smoke has curled up to spread the news far and wide. On leaving some place we have said to each other, 'Thank heaven, we have at last fairly left these wretches!' when one more faint halloo from an all-powerful voice, heard at a prodigious distance, would reach our ears, and clearly could we distinguish 'yammer-schooner'. But now, the more Fuegians the merrier; and very merry work it was. Both parties laughing, wondering, gaping at each other; we pitying them for giving us good fish and crabs for rags, etc.; they grasping at the chance of finding people so foolish as to exchange such splendid ornaments for a good supper. It was most amusing to see the undisguised smile of satisfaction with which one young woman, with her face painted black, tied several bits of scarlet cloth round her head with rushes. Her husband, who enjoyed the very universal privilege in this country of possessing two wives, evidently became jealous of all the attention paid to his young wife, and, after a consultation with his naked beauties, was paddled away by them.

Some of the Fuegians plainly showed that they had a fair notion of barter. I gave one man a large nail (a most valuable present) without making any signs for a return; but he immediately picked out two fish, and handed them up on the point of his spear. If any present was designed for one canoe, and it fell near another, it was invariably given to the right owner. The Fuegian boy, whom Mr Low had on board, showed, by going into the most violent passion, that he quite understood the reproach of being called a liar, which in truth he was. We were this time, as on all former occasions, much surprised at the little notice, or rather none whatever, which was taken of many things, the use of which must have been evident to the natives. Some circumstances – such as the beauty of scarlet cloth or blue beads, the absence of women, our care in washing ourselves – excited their admiration far more than any grand or complicated object, such as our ship. Bougainville has well remarked concerning these people, that they treat the 'chef-d'oeuvres de l'industrie humaine, comme ils traitent les loix de la nature et ses phénomènes'.

On the 5th of March we anchored in the cove at Woollya, but we saw not a soul there. We were alarmed at this, for the natives in Ponsonby Sound showed by gestures that there had been fighting; and we afterwards heard that the dreaded Oens men had made a descent. Soon a canoe, with a little flag flying, was seen approaching, with one of the men in it washing the paint off his face. This man was poor Jemmy – now a thin haggard savage, with long disordered hair, and naked, except a bit of a blanket round his waist. We did not recognize him till he was close to us; for he was ashamed of himself, and turned his back to the ship. We had left him plump, fat, clean, and well dressed; I never saw so complete and grievous a change. As soon, however, as he was clothed, and the first flurry was over, things wore a good appearance. He dined with Captain Roy, and ate his dinner as tidily as formerly. He told us he had 'too much' (meaning enough) to eat, that he was not cold, that his relations were very good people, and that he did not wish to go back to England.

10
The Milky Way

Joshua Slocum was the first man to sail round the world single-handed. This was between 1895 and 1898. His ship, the *Spray*, he practically built himself, having been given an old hulk which he took to pieces bit by bit, replacing each part as he went along with timber sawn from near-by trees which he had felled. *Spray* was 36 ft overall, 14 ft wide and 4 ft 2 in. in the hold; she was thus a beamy, shallow-draught boat.

Slocum was the son of a Nova Scotian farmer, but ran away to sea at the age of twelve, thus beginning his life as a professional sailor. Sometimes he sailed his own ships, but he was not particularly successful in his various enterprises.

I suppose that, if he had continued in this way, no one would ever have heard of him, except a few of his friends; but then, at the age of fifty-one, he set out on his voyage round the world in *Spray*. I cannot read his account without marvelling at his competence as a navigator, as he had no chronometer and his cheap clock lost its minute hand during the voyage. Like Darwin some sixty years before, he met the Fuegians.

A. R.

from '*SAILING ALONE AROUND THE WORLD*'
by JOSHUA SLOCUM

IT was the 3rd of March when the *Spray* sailed from Port Tamar direct for Cape Pillar[1], with the wind from the north-east, which I fervently hoped might hold till she cleared the land; but there was no such good luck in store. It soon began to rain and thicken in the north-west, boding no good. The *Spray* neared Cape Pillar rapidly, and, nothing loath, plunged into the Pacific Ocean at once, taking her first bath of it in the

---

[1] A headland 400 miles from The Horn up the Pacific coast of South America.

gathering storm. There was no turning back even had I wished to do so, for the land was now shut out by the darkness of night. The wind freshened, and I took in a third reef. The sea was confused and treacherous. In such a time as this the old fisherman prayed, 'Remember, Lord, my ship is so small and thy sea is so wide!' I saw now only the gleaming crests of waves. They showed white teeth while the sloop balanced over them. 'Everything for an offing,' I cried, and to this end I carried on all the sail she would bear. She ran all night with a free sheet, but on the morning of March 4th the wind shifted to southwest, then back suddenly to north-west, and blew with terrific force. The *Spray*, stripped of her sails, then bore off under bare poles. No ship in the world could have stood up against so violent a gale. Knowing that this storm might continue for many days, and that it would be impossible to work back to the westward along the coast outside of Tierra del Fuego, there seemed nothing to do but to keep on and go east about, after all. Anyhow, for my present safety the only course lay in keeping her before the wind. And so she drove south-east, as though about to round the Horn, while the waves rose and fell and bellowed their never-ending story of the sea; but the Hand that held these held also the *Spray*. She was running now with a reefed forestay-sail, the sheets flat amidship. I paid out two long ropes to steady her course and to break combing seas astern, and I lashed the helm amidship. In this trim she ran before it, shipping never a sea. Even while the storm raged at its worst, my ship was wholesome and noble. My mind as to her seaworthiness was put to ease for aye.

When all had been done that I could do for the safety of the vessel, I got to the fore-scuttle, between seas, and prepared a pot of coffee over a wood fire, and made a good Irish stew. Then, as before and afterwards on the *Spray*, I insisted on warm meals. In the tide-race off Cape Pillar, however, where the sea was marvellously high, uneven, and crooked, my appetite was slim, and for a time I postponed cooking. (Confidentially, I was seasick!)

The first day of the storm gave the *Spray* her actual test in the worst sea that Cape Horn or its wild regions could afford, and in no part of the world could a rougher sea be found than at this particular point, namely, off Cape Pillar, the grim sentinel of the Horn.

Farther offshore, while the sea was majestic, there was less apprehension of danger. There the *Spray* rode, now like a bird on the crest of a wave, and now like a waif deep down in the hollow between seas; and so she drove on. Whole days passed, counted as other days, but with always a thrill – yes, of delight.

On the fourth day of the gale, rapidly nearing the pitch of Cape Horn, I inspected my chart and pricked off the course and distance to Port Stanley, in the Falkland Islands, where I might find my way and refit, when I saw through a rift in the clouds a high mountain, about seven leagues away on the port beam. The fierce edge of the gale by this time had blown off, and I had already bent a square-sail on the boom in place of the mainsail, which was torn to rags. I hauled in the trailing ropes, hoisted this awkward sail reefed, the forestay-sail being already set, and under this sail brought her at once on the wind heading for the land, which appeared as an island in the sea. So it turned out to be, though not the one I had supposed.

I was exultant over the prospect of once more entering the Strait of Magellan and beating through again into the Pacific, for it was more than rough on the outside coast of Tierra del Fuego. It was indeed a mountainous sea. When the sloop was in the fiercest squalls, with only the reefed forestay-sail set, even that small sail shook her from keelson to truck when it shivered by the leech. Had I harboured the shadow of a doubt for her safety, it would have been that she might spring a leak in the garboard at the heel of the mast; but she never called me once to the pump. Under pressure of the smallest sail I could set she made for the land like a race-horse, and steering her over the crests of the waves so that she might not trip was nice work. I stood at the helm now and made the most of it.

Night closed in before the sloop reached the land, leaving her feeling the way in pitchy darkness. I saw breakers ahead before long. At this I wore ship and stood offshore, but was immediately startled by the tremendous roaring of breakers again ahead and on the lee bow. This puzzled me, for there should have been no broken water where I supposed myself to be. I kept off a good bit, then wore round, but finding broken water also there, threw her head again offshore. In this way, among dangers, I spent the rest of the night. Hail and sleet in the fierce squalls cut my flesh till the blood trickled over my face; but what of that? It was daylight, and the sloop was in the midst of the Milky Way of the sea, which is north-west of Cape Horn, and it was the white breakers of a huge sea over sunken rocks which had threatened to engulf her through the night. It was Fury Island I had sighted and steered for, and what a panorama was before me now and all around! It was not the time to complain of a broken skin. What could I do but fill away among the breakers and find a channel between them, now that it was day? Since she had escaped the rocks through the night, surely she would find her way by daylight. This was the greatest sea adventure of my life. God knows how my vessel escaped.

The sloop at last reached inside of small islands that sheltered her in smooth water. Then I climbed the mast to survey the wild scene astern. The great naturalist Darwin looked over this seascape from the deck of the *Beagle*, and wrote in his journal, 'Any landsman seeing the Milky Way would have nightmare for a week.' He might have added 'or seaman' as well.

The *Spray*'s good luck followed fast. I discovered, as she sailed along through a labyrinth of islands, that she was in the Cockburn Channel, which leads into the Strait of Magellan at a point opposite Cape Froward, and that she was already passing Thieves' Bay, suggestively named. And at night, March 8th, behold, she was at anchor in a snug cove at the Turn! Every heartbeat on the *Spray* now counted thanks.

Here I pondered on the events of the last few days, and, strangely enough, instead of feeling rested from sitting or lying down, I now began to feel jaded and worn; but a hot meal of venison stew soon put me right, so that I could sleep. As drowsiness came on I sprinkled the deck with tacks, and then I turned in, bearing in mind the advice of my old friend Samblich that I was not to step on them myself. I saw to it that not a few of them stood 'business end' up; for when the *Spray* passed Thieves' Bay two canoes had put out and followed in her wake, and there was no disguising the fact any longer that I was alone.

Now, it is well known that one cannot step on a tack without saying something about it. A pretty good Christian will whistle when he steps on the 'commercial end' of a carpet-tack; a savage will howl and claw the air, and that was just what happened that night about twelve o'clock, while I was asleep in the cabin, where the savages thought they 'had me', sloop and all, but changed their minds when they stepped on deck, for then they thought that I or somebody else had them. I had no need of a dog; they howled like a pack of hounds. I had hardly use for a gun. They jumped pell-mell, some into their canoes and some into the sea, to cool off, I suppose, and there was a deal of free language over it as they went. I fired several guns when I came on deck, to let the rascals know that I was home, and then I turned in again, feeling sure I should not be disturbed any more by people who left in so great a hurry.

The Fuegians, being cruel, are naturally cowards; they regard a rifle with superstitious fear. The only real danger one could see that might come from their quarter would be from allowing them to surround one within bow-shot, or to anchor within range where they might lie in ambush. As for their coming on deck at night, even had I not put tacks about, I could have cleared them off by shots from the cabin and hold. I always kept a quantity of ammunition within reach in the hold and in the cabin and in the forepeak, so that retreating to any of these places I could 'hold the fort' simply by shooting up through the deck.

Perhaps the greatest danger to be apprehended was from the use of fire. Every canoe carries fire; nothing is thought of that, for it is their custom to communicate by smoke-signals. The harmless brand that lies smouldering in the bottom of one of their canoes might be ablaze in one's cabin if he were not on the alert. The port captain of Sandy Point warned me particularly of this danger. Only a short time before they had fired a Chilean gunboat by throwing brands in through the stern windows of the cabin. The *Spray* had no openings in the cabin or deck, except two scuttles, and these were guarded by fastenings which could not be undone without waking me if I were asleep.

On the morning of the 9th, after a refreshing rest and a warm breakfast, and after I had swept the deck of tacks, I got out what spare canvas there was on board, and began to sew the pieces together in the shape of a peak for my square-mainsail, the tarpaulin. The day to all appearances promised fine weather and light winds, but appearances in Tierra del Fuego do not always count. While I was wondering why no trees grew on the slope abreast of the anchorage, half minded to lay by the sail-making and land with my gun for some game and to inspect a white boulder on the beach, near the brook, a williwaw came down with such terrific force as to carry the *Spray*, with two anchors down, like a feather out of the cove and away into deep water. No wonder trees did not grow on the side of that hill! Great Boreas! a tree would need to be all roots to hold on against such a furious wind.

From the cove to the nearest land to leeward was a long drift, however, and I had ample time to weigh both anchors before the sloop came near any danger, and so no harm came of it. I saw no more savages that day or the next; they probably had some sign by which they knew of the coming williwaws; at least, they were wise in not being afloat even on the second day, for I had no sooner gotten to work at sail-making again, after the anchor was down, than the wind, as on the day before, picked the sloop up and flung her seaward with a vengeance, anchor and all, as before. This fierce wind, usual to the Magellan country, continued on through the day, and swept the sloop by several miles of steep bluffs and precipices overhanging a bold shore of wild and uninviting appearance. I was not sorry to get away from it, though in doing so it was no Elysian shore to which I shaped my course. I kept on sailing in hope, since I had no choice but to go on, heading across for St Nicholas Bay, where I had cast anchor February 19th. It was now the 10th of March! Upon reaching the bay the second time I had circumnavigated the wildest part of desolate Tierra del Fuego. But the *Spray* had not yet arrived at St Nicholas, and by the merest accident her bones were saved from resting there when she did arrive. The parting of a staysail-sheet in a williwaw, when the sea was turbulent and she was plunging into the storm, brought me forward to see instantly a dark cliff ahead and breakers so close under the bows that I felt surely lost, and in my thoughts cried, 'Is the hand of fate against me, after all, leading me in the end to this dark spot?' I sprang aft again, unheeding the flapping sail, and threw the wheel over, expecting, as the sloop came down into the hollow of a wave, to feel her timbers smash under me on the rocks. But at the touch of her helm she swung clear of the danger, and in the next moment she was in the lee of the land.

It was the small island in the middle of the bay for which the sloop had been steering, and which she made with such unerring aim as nearly to run it down. Farther along in the bay was the anchorage, which I managed to reach, but before I could get the anchor down another squall caught the sloop and whirled her round like a top and carried her away, altogether to leeward of the bay. Still farther to leeward was a great headland, and I bore off for that. This was retracing my course towards Sandy Point, for the gale was from the south-west.

I had the sloop soon under good control, however, and in a short time rounded to under the lee of a mountain, where the sea was as smooth as a mill-pond, and the sails flapped and hung limp while she carried her way close in. Here I thought I would anchor and rest till morning, the depth being eight fathoms very close to the shore. But it was interesting to see, as I let go the anchor, that it did not reach the bottom before another williwaw struck down from this mountain and carried the sloop off faster than I could pay out cable. Therefore, instead of resting, I had to 'man the windlass' and heave up the anchor with fifty fathoms of cable hanging up and down in deep water. This was in that part of the strait called Famine Reach. Dismal Famine Reach! On the sloop's crab-windlass I worked the rest of the night, thinking how much easier it was for me when I could say, 'Do that thing or the other,' than now doing all myself. But I hove away and sang the old chants that I sang when I was a sailor. Within the last few days I had passed through much and was now thankful that my state was no worse.

It was daybreak when the anchor was at the hawse. By this time the wind had gone down, and cat's-paws took the place of williwaws, while the sloop drifted slowly towards Sandy Point. She came within sight of ships at anchor in the roads, and I was more than half minded to put in for new sails, but the wind coming out from the north-east, which was fair for the other direction, I turned the prow of the *Spray* westward once more for the Pacific, to traverse a second time the second half of my first course through the strait.

II

A Day of Great Disaster

A great deal of literature describes the events of Scott's last expedition to the Antarctic, but I always find Scott's own writing in his journal quite the most dramatic. His tragic account of his journey to the Pole is very well known, but few remember that a storm nearly ended the whole expedition before it even reached the Antarctic.

Many a yachtsman has been in trouble when the suction pipes to their bilge pumps get choked, so they would understand what this meant in a ship heavily laden with all the equipment for a long expedition.

A. R.

from '*SCOTT'S LAST EXPEDITION*'
by CAPTAIN R. F. SCOTT, R.N., C.V.O.

FRIDAY, *December 2nd.* A day of great disaster. From four o'clock last night the wind freshened with great rapidity, and very shortly we were under topsails, jib, and staysail only. It blew very hard and the sea got up at once. Soon we were plunging heavily and taking much water over the lee rail. Oates and Atkinson with intermittent assistance from others were busy keeping the ponies on their legs. Cases of petrol, forage, etc., began to break loose on the upper deck; the principal trouble was caused by the loose coal-bags, which were bodily lifted by the seas and swung against the lashed cases. 'You know how carefully everything had been lashed, but no lashings could have withstood the onslaught of these coal-sacks for long'; they acted like battering rams. 'There was nothing for it but to grapple with the evil, and nearly all hands were

Note:—Passages enclosed in inverted commas are taken from Scott's letters home.

labouring for hours in the waist of the ship, heaving coal sacks overboard and re-lashing the petrol cases, etc., in the best manner possible under such difficult and dangerous circumstances. The seas were continually breaking over these people and now and again they would be completely submerged. At such times they had to cling for dear life to some fixture to prevent themselves being washed overboard, and with coalbags and loose cases washing about, there was every risk of such hold being torn away.

'No sooner was some semblance of order restored than some exceptionally heavy wave would tear away the lashing and the work had to be done all over again.'

The night wore on, the sea and wind ever rising, and the ship ever plunging more distractedly; we shortened sail to main topsail and staysail, stopped engines and hove to, but to little purpose. Tales of ponies down came frequently from forward, where Oates and Atkinson laboured through the entire night. Worse was to follow, much worse – a report from the engine-room that the pumps had choked and the water risen over the gratings.

From this moment, about 4 a.m., the engine-room became the centre of interest. The water gained in spite of every effort. Lashly, to his neck in rushing water, stuck gamely to the work of clearing suctions. For a time, with donkey engine and bilge pump sucking, it looked as though the water would be got under; but the hope was short-lived: five minutes of pumping invariably led to the same result – a general choking of the pumps.

The outlook appeared grim. The amount of water which was being made, with the ship so roughly handled, was most uncertain. 'We knew that normally the ship was not making much water, but we also knew that a considerable part of the water washing over the upper deck must be finding its way below; the decks were leaking in streams. The ship was very deeply laden; it did not need the addition of much water to get her waterlogged, in which condition anything might have happened.'

The hand-pump produced only a dribble, and its suction could not be got at; as the water crept higher it got in contact with the boiler and grew warmer – so hot at last that no one could work at the suctions. Williams had to confess he was beaten and must draw fires. What was to be done? Things for the moment appeared very black. The sea seemed higher than ever; it came over lee rail and poop, a rush of green water; the ship wallowed in it; a great piece of the bulwark carried clean away. The bilge pump is dependent on the main engine. To use the pump it was necessary to go ahead. It was at such times that the heaviest seas swept in over the lee rail; over and over again the rail, from the forerigging to the main, was covered by a solid sheet of curling water which swept aft and high on the poop. On one occasion I was waist deep when standing on the rail of the poop.

The scene on deck was devastating, and in the engine-room the water, though really not great in quantity, rushed over the floor plates and frames in a fashion that gave it a fearful significance.

The afterguard were organized in two parties by Evans to work buckets; the men were kept steadily going on the choked hand-pumps – this seemed all that could be done for the moment, and what a measure to count as the sole safeguard of the ship from sinking, practically an attempt to bale her out! Yet strange as it may seem the effort has not been wholly fruitless – the string of buckets which has now been kept going for four hours, together with the dribble from the pump, has kept the water under – if anything there is a small decrease.

Meanwhile we have been thinking of a way to get at the suction of the pump: a hole is being made in the engine-room bulkhead, the coal between this and the pump shaft will be removed, and a hole made in the shaft. With so much water coming on board, it is impossible to open the hatch over the shaft. We are not out of the wood, but hope dawns, as indeed it should for me, when I find myself so wonderfully served. Officers and men are singing chanties over their arduous work. Williams

is working in sweltering heat behind the boiler to get the door
made in the bulkhead. Not a single one has lost his good spirits.
A dog was drowned last night, one pony is dead and two others
in a bad condition – probably they too will go. 'Occasionally
a heavy sea would bear one of them away, and he was only
saved by his chain. Meares with some helpers had constantly to
be rescuing these wretched creatures from hanging, and trying to
find them better shelter, an almost hopeless task. One poor
beast was found hanging when dead; one was washed away
with such force that his chain broke and he disappeared overboard; the next wave miraculously washed him onboard again
and he is now fit and well.' The gale has exacted heavy toll,
but I feel all will be well if we can only cope with the water.
Another dog has just been washed overboard – alas! Thank
God, the gale is abating. The sea is still mountainously high,
but the ship is not labouring so heavily as she was. I pray we
may be under sail again before morning.

Saturday, December 3rd. Yesterday the wind slowly fell towards
evening; less water was taken on board, therefore less found its
way below, and it soon became evident that our baling was
gaining on the engine-room. The work was steadily kept going
in two-hour shifts. By 10 p.m. the hole in the engine-room bulkhead was completed, and (Lieut.) Evans, wriggling over the
coal, found his way to the pump shaft and down it. He soon
cleared the suction 'of the coal balls (a mixture of coal and oil)
which choked it', and to the joy of all a good stream of water
came from the pump for the first time. From this moment it was
evident we should get over the difficulty, and though the pump
choked again on several occasions the water in the engine-room
steadily decreased. It was good to visit that spot this morning
and to find that the water no longer swished from side to side.
In the forenoon fires were laid and lighted – the hand-pump
was got into complete order and sucked the bilges almost dry,
so that great quantities of coal and ashes could be taken out.

12
The Voyage of the James Caird

When the *Endurance* was crushed in the Antarctic ice and sank, Shackleton and his men made their way to Elephant Island. They managed to drag with them two of the boats from the *Endurance*, besides some provisions. Soon shortage of food and the approach of the Antarctic winter made their position serious, so Shackleton decided to set out with five others in one of the open boats, to get help for the rest from a whaling settlement over eight hundred miles away across the Southern Ocean.

A. R.

from *'SOUTH'* by SIR ERNEST SHACKLETON

I DISCUSSED with Wild and Worsley the chances of reaching South Georgia before the winter locked the seas against us. Some effort had to be made to secure relief. Privation and exposure had left their mark on the party, and the health and mental condition of several men were causing me serious anxiety. Then the food-supply was a vital consideration. We had left ten cases of provisions in the crevice of the rocks at our first camping-place on the island. An examination of our stores showed that we had full rations for the whole party for a period of five weeks. The rations could be spread over three months on a reduced allowance and probably would be supplemented by seals and sea-elephants to some extent. I did not dare to count with full confidence on supplies of meat and blubber, for the animals seemed to have deserted the beach and the winter was near. Our stocks included three seals and two and a half skins (with blubber attached). We were mainly dependent on the blubber for fuel, and, after making a preliminary survey of the situation, I decided that the party must be limited to one hot meal a day.

A boat journey in search of relief was necessary and must not be delayed. That conclusion was forced upon me. The nearest port where assistance could certainly be secured was Port Stanley, in the Falkland Islands, 540 miles away, but we could scarcely hope to beat up against the prevailing north-westerly wind in a frail and weakened boat with a small sail area. South Georgia was over 800 miles away, but lay in the area of the west winds, and I could count upon finding whalers at any of the whaling-stations on the east coast. A boat party might make the voyage and be back with relief within a month, provided that the sea was clear of ice and the boat survive the great seas. It was not difficult to decide that South Georgia must be the objective, and I proceeded to plan ways and means. The hazards of a boat journey across 800 miles of stormy sub-Antarctic ocean were obvious, but I calculated that at worst the venture would add nothing to the risks of the men left on the island. There would be fewer mouths to feed during the winter and the boat would not require to take more than one month's provisions for six men, for if we did not make South Georgia in that time we were sure to go under. A consideration that had weight with me was that there was no chance at all of any search being made for us on Elephant Island.

The case required to be argued in some detail, since all hands knew that perils of the proposed journey were extreme. The risk was justified solely by our urgent need of assistance. The ocean south of Cape Horn in the middle of May is known to be the most tempestuous storm-swept area of water in the world. The weather then is unsettled, the skies are dull and overcast, and the gales are almost unceasing. We had to face these conditions in a small and weather-beaten boat, already strained by the work of the months that had passed. Worsley and Wild realized that the attempt must be made, and they both asked to be allowed to accompany me on the voyage. I told Wild at once that he would have to stay behind. I relied upon him to hold the party together while I was away, and to make the best of his way to Deception Island with the men in the spring in the event of our failure to bring help. Worsley I would take with me, for I had a very high opinion of his accuracy and quickness as navigator, and especially in the snapping and working out of positions in difficult circumstances – an opinion that was only enhanced during the actual journey. Four other men would be required and I decided to call for volunteers, although, as a matter of fact, I pretty well knew which of the people I would select. Crean I proposed to leave on the island as a right-hand man for Wild, but he begged so hard to be allowed to come in the boat that, after consultation with Wild, I promised to take him. I called the men together, explained my plan, and asked for volunteers. Many came forward at once. Some were not fit enough for the work that would have to be done, and others would not have been much use in the boat since they were not seasoned sailors, though the experiences of recent months entitled them to some consideration as seafaring men. I finally selected McNeish, McCarthy, and Vincent in addition to Worsley and Crean. The crew seemed a strong one, and as I looked at the men I felt confidence increasing.

The decision made, I walked through the blizzard with Worsley and Wild to examine the *James Caird*. The 20-ft boat had never looked big; she appeared to have shrunk in some mysterious way when I viewed her in the light of our new undertaking. She was an ordinary ship's whaler, fairly strong, but showing signs of the strains she had endured since the crushing of the *Endurance*. Where she was holed in leaving the pack was, fortunately, about the water-line and easily patched. Standing beside her, we glanced at the fringe of the storm-swept, tumultuous sea that formed our path. Clearly, our voyage would be a big adventure. I called the carpenter and asked him if he could do anything to make the boat more seaworthy. He first inquired if he was to go with me, and seemed quite pleased when I said 'Yes.' He was over fifty years of age, and not altogether fit, but he had a good knowledge of sailing boats and was very quick.

McCarthy said that he could contrive some sort of covering for the *James Caird* if he might use the lids of the cases and the four sledge-runners that we had lashed inside the boat for use in the event of a landing on Graham Land at Wilhelmina Bay. This bay, at one time the goal of our desire, had been left behind in the course of our drift, but we had retained the runners. The carpenter proposed to complete the covering with some of our canvas, and he set about making his plans at once.

Noon had passed and the gale was more severe than ever. We could not proceed with our preparations that day. The tents were suffering in the wind and the sea was rising. The gale was stronger than ever on the following morning (April 20th). No work could be done. Blizzard and snow, snow and blizzard, sudden lulls and fierce returns. During the lulls we could see on the far horizon to the north-east bergs of all shapes and sizes driving along before the gale, and the sinister appearance of the swift-moving masses made us thankful indeed that instead of battling with the storm amid the ice, we were required only to face the drift from the glaciers and the inland heights. The gusts might throw us off our feet, but at least we fell on solid ground and not on rocking floes.

There was a lull in the bad weather on April 21st, and the carpenter started to collect material for the decking of the *James Caird*. He fitted the mast of the *Stancomb Wills* fore and aft inside the *James Caird* as a hog-back, and thus strengthened the keel with the object of preventing our boat 'hogging' – that is, buckling in heavy seas. He had not sufficient wood to provide a deck, but by using the sledge-runners and box-lids he made a framework extending from the forecastle aft to a well. It was a patched-up affair, but it provided a base for a canvas covering. He had a bolt of canvas frozen stiff, and this material had to be cut and then thawed out over the blubber-stove, foot by foot, in order that it might be sewn into the form of a cover. When it had been nailed and screwed into position it certainly gave an appearance of safety to the boat, though I had an uneasy feeling that it bore a strong likeness to stage scenery, which may look like a granite wall and is in fact nothing better than canvas and lath. As events proved, the covering served its purpose well. We certainly could not have lived through the voyage without it.

Another fierce gale was blowing on April 22nd, interfering with our preparations for the voyage. We were setting aside stores for the boat journey and choosing the essential equipment from the scanty stock at our disposal. Two ten-gallon casks had to be filled with water melted down from ice collected at the foot of the glacier. This was rather a slow business.

The weather was fine on April 23rd, and we hurried forward our preparations. It was on this day I decided finally that the crew for the *James Caird* should consist of Worsley, Crean, McNeish, McCarthy, Vincent, and myself. A storm came on about noon, with driving snow and heavy squalls. Occasionally the air would clear for a few minutes, and we could see a line of pack-ice, five miles out, driving across from west to east. The sight increased my anxiety to get away quickly. Winter was advancing, and soon the pack might close completely round the island and stay our departure for days, or even weeks.

Worsley, Wild, and I climbed to the summit of the seaward rocks and examined the ice from a better vantage-point than the beach offered. The belt of pack outside appeared to be sufficiently broken for our purposes, and I decided that, unless the conditions forbade it, we would make a start in the *James Caird* on the following morning. Obviously the pack might close at any time. This decision made, I spent the rest of the day looking over the boat, gear, and stores, and discussing plans with Worsley and Wild.

Our last night on the solid ground of Elephant Island was cold and uncomfortable. We turned out at dawn and had breakfast. Then we launched the *Stancomb Wills* and loaded her with stores, gear, and ballast, which would be transferred to the *James Caird* when the heavier boat had been launched. The ballast consisted of bags made from blankets and filled with sand, making a total weight of about 1000 lb. In addition we had gathered a number of round boulders and about 250 lb. of ice, which would supplement our two casks of water.

The swell was slight when the *Stancomb Wills* was launched and the boat got under way without any difficulty; but half an hour later, when we were pulling down the *James Caird*, the swell increased suddenly. Apparently the movement of the ice outside had made an opening and allowed the sea to run in without being blanketed by the line of pack. The swell made things difficult. Many of us got wet to the waist while dragging the boat out – a serious matter in that climate. When the *James Caird* was afloat in the surf she nearly capsized among the rocks before we could get her clear, and Vincent and the carpenter, who were on the deck, were thrown into the water. This was really bad luck for the two men would have small chance of drying their clothes after we had got under way.

The *James Caird* was soon clear of the breakers. We used all the available ropes as a long painter to prevent her drifting away to the north-east, and then the *Stancomb Wills* came along-side, transferred her load, and went back to the shore for more.

By midday the *James Caird* was ready for the voyage. Vincent and the carpenter had secured some dry clothes by exchange with members of the shore party (I heard afterwards that it was a full fortnight before the soaked garments were finally dried), and the boat's crew was standing by waiting for the order to cast off. A moderate westerly breeze was blowing, I went ashore in the *Stancomb Wills* and had a last word with Wild, who was remaining in full command, with directions as to his course of action in the event of our failure to bring relief, but I practically left the whole situation and scope of action and decision to his own judgement, secure in the knowledge that he would act wisely. I told him that I trusted the party to him, and said good-bye to the men. Then we pushed off for the last time, and within a few minutes I was aboard the *James Caird*. The crew of the *Stancomb Wills* shook hands with us as the boats bumped together and offered us the last good wishes. Then, setting our jib, we cut the painter and moved away to the north-east. The men who were staying behind made a pathetic little group on the beach, with the grim heights of the island behind them and the sea seething at their feet, but they waved to us and gave three hearty cheers. There was hope in their hearts and they trusted us to bring the help that they needed.

I had all sails set, and the *James Caird* quickly dipped the beach and its line of dark figures. The westerly wind took us rapidly to the line of pack, and as we entered it I stood up with my arm around the mast, directing the steering, so as to avoid the great lumps of ice that were flung about in the heave of the sea. The pack thickened and we were forced to turn almost due east, running before the wind towards a gap I had seen in the morning from the high ground. I could not see the gap now, but we had come out on its bearing and I was prepared to find that it had been influenced by the easterly drift. At four o'clock in the afternoon we found the channel, much narrower than it had seemed in the morning but still navigable. Dropping sail we rowed through without touching the ice anywhere, and by 5.30 p.m. we were clear of the pack with open water before us. We passed one more piece of ice in the darkness an hour later, but the pack lay behind, and with a fair wind swelling the sails we steered our little craft through the night, our hopes centred on our distant goal. The swell was very heavy now, and when the time came for our first evening meal we found great difficulty in keeping the Primus lamp alight and preventing the hoosh splashing out of the pot. Three men were needed to attend to the cooking, one man holding the lamp and two men guarding the aluminium cooking-pot, which had to be lifted clear of the Primus whenever the movement of the boat threatened to cause a disaster. Then the lamp had to be protected from water, for sprays were coming over the bows and our flimsy decking was by no means watertight. All these operations were conducted in the confined space under the decking, where the men lay or knelt and adjusted themselves as best they could to the angles of our cases and ballast. It was uncomfortable, but we found

consolation in the reflection that without the decking we could
not have used the cooker at all.

The tale of the next sixteen days is one of supreme strife amid
heaving waters. The sub-Antarctic Ocean lived up to its evil
winter reputation. I decided to run north for at least two days
while the wind held and so get into warmer weather before
turning to the east and laying a course for South Georgia. We
took two-hourly spells at the tiller. The men who were not on
watch crawled into the sodden sleeping-bags and tried to
forget their troubles for a period; but there was no comfort in
the boat. The bags and cases seemed to be alive in the unfailing
knack of presenting their most uncomfortable angles to our rest-
seeking bodies. A man might imagine for a moment that he had
found a position of ease, but always discovered quickly that
some unyielding point was impinging on muscle or bone. The
first night aboard the boat was one of acute discomfort for us all,
and we were heartily glad when the dawn came and we could
set about the preparation of a hot breakfast.

By running north for the first two days I hoped to get warmer
weather and also to avoid lines of pack that might be extending
beyond the main body. We needed all the advantage that we
could obtain from the higher latitude for sailing on the great
circle, but we had to be cautious regarding possible ice streams.
Cramped in our narrow quarters and continually wet by the
spray, we suffered severely from cold throughout the journey.
We fought the seas and the winds, and at the same time had a
daily struggle to keep ourselves alive. At times we were in dire
peril. Generally we were upheld by the knowledge that we were
making progress towards the land where we would be, but
there were days and nights when we lay hove to, drifting across
the storm-whitened seas and watching, with eyes interested
rather than apprehensive, the uprearing masses of water, flung
to and fro by Nature in the pride of her strength. Deep seemed
the valleys when we lay between the reeling seas. High were the
hills when we perched momentarily on the top of giant com-
bers. Nearly always there were gales. So small was our boat and

so great were the seas that often our sail flapped idly in the calm
between the crests of two waves. Then we would climb the next
slope and catch the full fury of the gale where the wool-like
whiteness of the breaking water surged around us.

The wind came up strong and worked into a gale from the
north-west on the third day out. We stood away to the east.
The increasing seas discovered the weakness of our decking.
The continuous blows shifted the box-lids and sledge-runners so
that the canvas sagged down and accumulated water. Then icy
trickles, distinct from the driving sprays, poured fore and aft
into the boat. The nails that the carpenter had extracted from
cases at Elephant Island and used to fasten down the battens
were too short to make firm the decking. We did what we could
to secure it, but our means were very limited, and the water
continued to enter the boat at a dozen points. Much bailing
was necessary, and nothing that we could do prevented our
gear from becoming sodden. The searching runnels from the
canvas were really more unpleasant than the sudden definite
douches of the sprays. Lying under the thwarts during watches
below, we tried vainly to avoid them. There were no dry places
in the boat, and at last we simply covered our heads with our
Burberrys and endured the all-pervading water. The bailing
was work for the watch. Real rest we had none. The perpetual
motion of the boat made repose impossible; we were cold, sore,
and anxious. We moved on hands and knees in the semi-dark-
ness of the day under the decking. The darkness was complete
by 6 p.m., and not until 7 a.m. of the following day could we see
one another under the thwarts. We had a few scraps of candle,
and they were preserved carefully in order that we might have
light at mealtimes. There was one fairly dry spot in the boat,
under the solid original decking at the bows, and we managed to
protect some of our biscuit from the salt water; but I do not
think any of us got the taste of salt out of our mouths during the
voyage.

The difficulty of movement in the boat would have had its
humorous side if it had not involved us in so many aches and

pains. We had to crawl under the thwarts in order to move along the boat, and our knees suffered considerably. When a watch turned out it was necessary for me to direct each man by name when and where to move, since if all hands had crawled about at the same time the result would have been dire confusion and many bruises. Then there was the trim of the boat to be considered. The order of the watch was four hours on and four hours off, three men to the watch. One man had the tiller-ropes, the second man attended to the sail, and the third bailed for all he was worth. Sometimes when the water in the boat had been reduced to reasonable proportions, our pump could be used. This pump, which Hurley had made from the Flinders bar case of our ship's standard compass, was quite effective, though its capacity was not large. The man who was attending to the sail could pump into the big outer cooker, which was lifted and emptied overboard when filled. We had a device by which the water could go direct from the pump into the sea through a hole in the gunwale, but this hole had to be blocked at an early stage of the voyage, since we found that it admitted water when the boat rolled.

While a new watch was shivering in the wind and spray, the men who had been relieved groped hurriedly among the soaked sleeping-bags and tried to steal a little of the warmth created by the last occupants; but it was not always possible for us to find even this comfort when we went off watch. The boulders that we had taken aboard for ballast had to be shifted continually in order to trim the boat and give access to the pump, which became choked with hairs from the moulting sleeping-bags and finneskoe. The four reindeer-skin sleeping-bags shed their hair freely owing to the continuous wetting, and soon became quite bald in appearance. The moving of the boulders was very weary and painful work. We came to know every one of the stones by sight and touch, and I have vivid memories of their angular peculiarities even today. They might have been of considerable interest as geological specimens to a scientific man under happier conditions. As ballast they were useful. As weights to be moved about in cramped quarters they were simply appalling. They spared no portion of our poor bodies.

Our meals were regular in spite of the gales. Breakfast, at 8 a.m., consisted of a pannikin of hot hoosh made from Bovril sledging ration, two biscuits, and some lumps of sugar. Lunch came at 1 p.m., and comprised Bovril sledging ration, eaten raw, and a pannikin of hot milk for each man. Tea, at 5 p.m., had the same menu. Then during the night we had a hot drink, generally of milk. The meals were the bright beacons in those cold and stormy days. The glow of warmth and comfort produced by the food and drink made optimists of us all.

A severe south-westerly gale on the fourth day out forced us to heave to. I would have liked to have run before the wind, but the sea was very high and the *James Caird* was in danger of broaching to and swamping. The delay was vexatious, since up to that time we had been making sixty or seventy miles a day; good going with our limited sail area. We hove to under double-reefed mainsail and our little jigger, and waited for the gale to blow itself out. During that afternoon we saw bits of wreckage, the remains probably of some unfortunate vessel that had failed to weather the strong gales south of Cape Horn. The weather conditions did not improve, and on the fifth day out the gale was so fierce that we were compelled to take in the double-reefed mainsail and hoist our small jib instead. We put out a sea-anchor to keep the *James Caird*'s head up to the sea. This anchor consisted of a triangular canvas bag fastened to the end of the painter and allowed to stream out from the bows. The boat was high enough to catch the wind, and as she drifted to leeward the drag of the anchor kept her head to windward. Thus our boat took most of the seas more or less end on. Even then the crests of the waves often would curl right over us and we shipped a great deal of water, which necessitated unceasing bailing and pumping. Looking out abeam, we would see a hollow like a tunnel formed as the crest of a big wave toppled over on to the swelling body of water. A thousand times it appeared as though the *James Caird* must be engulfed; but the boat lived. The south-westerly gale had its birthplace above the Antarctic Continent and its freezing breath lowered the temperature far towards zero. The sprays froze upon the boat, and gave bows, sides, and decking a heavy coat of mail. This accumulation of ice reduced the buoyancy of the boat, and to that extent was an added peril; but it possessed a notable advantage from one point of view. The water ceased to drop and trickle from the canvas, and the spray came in solely at the well in the after part of the boat. We could not allow the load of ice to grow beyond a certain point, and in turns we crawled about the decking forward, chipping and picking at it with the available tools.

When daylight came on the morning of the sixth day out, we saw and felt that the *James Caird* had lost her resiliency. She was not rising to the oncoming seas. The weight of the ice that had formed in her and upon her during the night was having its effect, and she was becoming more like a log than a boat. The situation called for immediate action. We first broke away the spare oars, which were encased in ice and frozen to the sides of the boat, and threw them overboard. We retained two oars for use when we got inshore. Two of the fur sleeping-bags went over the side; they were thoroughly wet, weighing probably 40 lb each, and they had frozen stiff during the night. Three men constituted the watch below, and when a man went down it was better to turn into the wet bag just vacated by another man than to thaw out a frozen bag with the heat of his unfortunate body. We now had four bags, three in use, and one for emergency use in case a member of the party should break down permanently. The reduction of weight relieved the boat to some extent, and vigorous chipping and scraping did more. We had to be very careful not to put axe or knife through the frozen canvas of the decking as we crawled over it, but gradually we got rid of a lot of ice. The *James Caird* lifted to the endless waves as though she lived again.

About 11 a.m. the boat suddenly fell off into the trough of the sea. The painter had parted and the sea-anchor had gone. This was serious. The *James Caird* went away to leeward, and we had no chance at all of recovering the anchor and our valuable rope, which had been our only means of keeping the boat's head up to the seas without the risk of hoisting sail in a gale. Now we had to set the sail and trust to its holding. While the *James Caird* rolled heavily in the trough, we beat the frozen canvas until the bulk of the ice had cracked off it, and then hoisted it. The frozen gear worked protestingly, but after a struggle our little craft came up to the wind again, and we breathed more freely.

We held the boat up to the gale during that day, enduring as best we could discomforts that amounted to pain. The boat tossed interminably on the big waves under grey, threatening skies. Our thoughts did not embrace much more than the necessities of the hour. Every surge of the sea was an enemy to be watched and circumvented. We ate our scanty meals, treated our frost-bites, and hoped for the improved conditions that the morrow might bring. Night fell early, and in the lagging hours of darkness we were cheered by a change for the better in the weather. The wind dropped, the snow-squalls became less frequent, the sea moderated. When the morning of the seventh day dawned, there was not much wind. We shook the reef out of the sail and laid our course once more for South Georgia. The sun came out bright and clear, and presently Worsley got a snap for longitude. We hoped that the sky would remain clear until noon, so that we could get the latitude. We had been six days out without an observation, and our dead reckoning naturally was uncertain. The boat must have presented a strange appearance that morning. All hands basked in the sun. We hung our sleeping-bags to the mast and spread our socks and other gear all over the deck. Some of the ice had melted off the *James Caird* in the early morning after the gale began to slacken and dry patches were appearing in the decking. Porpoises came blowing round the boat, and Cape pigeons wheeled and swooped within a few feet of us. These little black-and-white birds have an air of friendliness that is not possessed by the great circling albatross. They had looked grey against the swaying sea during the storm as they darted about over our heads and uttered their plaintive cries. The albatrosses, of the black or sooty variety, had watched with hard, bright eyes, and seemed to have a quite impersonal interest in our struggle to keep afloat amid the battering seas. In addition to the Cape pigeons an occasional stormy petrel flashed overhead. Then there was a small bird, unknown to me, that appeared always to be in a fussy, bustling state, quite out of keeping with the surroundings. It irritated me. It had practically no tail, and it flitted about vaguely as though in search of the lost member. I used to find myself wishing it would find its tail and have done with the silly fluttering.

We revelled in the warmth of the sun that day. Life was not so bad, after all. We felt we were well on our way. Our gear was drying, and we could have a hot meal in comparative comfort. The swell was still heavy, but it was not breaking and the boat rode easily. At noon Worsley balanced himself on the gunwale and clung with one hand to the stay of the mainmast while he got a snap of the sun. The result was more than encouraging. We had done over 380 miles and were getting on for half-way to South Georgia. It looked as though we were going to get through.

The wind freshened to a good stiff breeze during the afternoon, and the *James Caird* made satisfactory progress. I had not realized until the sunlight came how small our boat really was. There was some influence in the light and warmth, some hint of happier days, that made us revive memories of other voyages, when we had stout decks beneath our feet, unlimited food at our command, and pleasant cabins for our ease. Now we clung to a battered little boat, 'alone, alone, all, all alone, alone on a wide, wide, sea'. So low in the water were we that each succeeding swell cut off our view of the skyline. We were a tiny speck in the vast vista of the sea – the ocean that is open to all, and merciful to none, that threatens even when it seems to yield, and that is pitiless always to weakness. For a moment the consciousness of the forces arrayed against us would be almost overwhelming. Then hope and confidence would rise again as our boat rose to a wave and tossed aside the crest in a sparkling shower like the play of prismatic colours at the foot of a waterfall. My double-barrelled gun and some cartridges had been stowed aboard the boat as an emergency precaution against a shortage of food, but we were not disposed to destroy our little neighbours, the Cape pigeons, even for the sake of fresh meat. We might have shot an albatross, but the wandering king of the ocean aroused in us something of the feeling that inspired, too late, the Ancient Mariner. So the gun remained among the stores and sleeping-bags in the narrow quarters beneath our leaking deck, and the birds followed us unmolested.

The eighth, ninth, and tenth days of the voyage had few features worthy of special note. The wind blew hard during those days, and the strain of navigating the boat was unceasing, but always we made some advance towards our goal. No bergs showed on our horizon, and we knew that we were clear of the ice-fields. Each day brought its little round of troubles, but also compensation in the form of food and growing hope. We felt that we were going to succeed. The odds against us had been great, but we were winning through. We still suffered severely from the cold, for, though the temperature was rising, our vitality was declining owing to shortage of food, exposure, and the necessity of maintaining our cramped positions day and night. I found that it was now absolutely necessary to prepare hot milk for all hands during the night, in order to sustain life till dawn. This meant lighting the Primus lamp in the darkness and involved an increased drain on our small store of matches. It was the rule that one match must serve when the Primus was being lit. We had no lamp for the compass, and during the early days of the voyage we would strike a match when the steersman wanted to see the course at night; but later the necessity for strict economy impressed itself upon us, and the practice of striking matches at night was stopped. We had one

watertight tin of matches. I had stowed away in a pocket, in
readiness for a sunny day, a lens from one of the telescopes, but
this was of no use during the voyage. The sun seldom shone
upon us. The glass of the compass got broken one night, and we
contrived to mend it with adhesive tape from the medicine-
chest. One of the memories that comes to me from those days is
of Crean singing at the tiller. He always sang while he was
steering, and nobody ever discovered what the song was. It was
devoid of tune and as monotonous as the chanting of a
Buddhist monk at his prayers; yet somehow it was cheerful. In
moments of inspiration Crean would attempt 'The Wearing of
the Green'.

On the tenth night Worsley could not straighten his body
after his spell at the tiller. He was thoroughly cramped, and we
had to drag him beneath the decking and massage him before he
could unbend himself and get into a sleeping-bag. A hard
north-westerly gale came up on the eleventh day (May 5th)
and shifted to the south-west in the late afternoon. The sky was
overcast and occasional snow squalls added to the discomfort
produced by a tremendous cross-sea – the worst, I thought, that
we had experienced. At midnight I was at the tiller and
suddenly noticed a line of clear sky between the south and south-
west. I called to the other men that the sky was clearing, and
then a moment later I realized that what I had seen was not a
rift in the clouds but the white crest of an enormous wave.
During twenty-six years' experience of the ocean in all its
moods I had not encountered a wave so gigantic. It was a
mighty upheaval of the ocean, a thing quite apart from the big
white-capped seas that had been our tireless enemies for many
days. I shouted, 'For God's sake, hold on! It's got us!' Then
came a moment of suspense that seemed drawn out into hours.
White surged the foam of the breaking sea around us. We felt
our boat lifted and flung forward like a cork in breaking surf.
We were in a seething chaos of tortured water; but somehow the
boat lived through it, half-full of water, sagging to the dead
weight and shuddering under the blow. We bailed with the

energy of men fighting for life, flinging the water over the sides with every receptacle that came to our hands, and after ten minutes of uncertainty we felt the boat renew her life beneath us. She floated again and ceased to lurch drunkenly as though dazed by the attack of the sea. Earnestly we hoped that never again would we encounter such a wave.

The conditions in the boat, uncomfortable before, had been made worse by the deluge of water. All our gear was thoroughly wet again. Our cooking-stove had been floating about in the bottom of the boat, and portions of our last hoosh seemed to have permeated everything. Not until 3 a.m., when we were all chilled almost to the limit of endurance, did we manage to get the stove alight and make ourselves hot drinks. The carpenter was suffering particularly, but he showed grit and spirit. Vincent had for the past week ceased to be an active member of the crew, and I could not easily account for his collapse. Physically he was one of the strongest men in the boat. He was a young man, he had served on North Sea trawlers, and he should have been able to bear hardships better than McCarthy, who, not so strong, was always happy.

The weather was better on the following day (May 6th), and we got a glimpse of the sun. Worsley's observation showed that we were not more than a hundred miles from the north-west corner of South Georgia. Two more days with a favourable wind and we would sight the promised land. I hoped that there would be no delay, for our supply of water was running very low. The hot drink at night was essential, but I decided that the daily allowance of water must be cut down to half a pint per man. The lumps of ice we had taken aboard had gone long ago. We were dependent upon the water we had brought from Elephant Island, and our thirst was increased by the fact that we were now using the brackish water in the breaker that had been slightly stove in in the surf when the boat was being loaded. Some sea-water had entered at that time.

Thirst took possession of us. I dared not permit the allowance of water to be increased since an unfavourable wind might drive us away from the island and lengthen our voyage by many days. Lack of water is always the most severe privation that men can be condemned to endure, and we found, as during our earlier boat voyage, that the salt water in our clothing and the salt spray that lashed our faces made our thirst grow quickly to a burning pain. I had to be very firm in refusing to allow anyone to anticipate the morrow's allowance, which I was sometimes begged to do. We did the necessary work dully and hoped for the land. I had altered the course to the east so as to make sure of our striking the island, which would have been impossible to regain if we had run past the northern end. The course was laid on our scrap of chart for a point some thirty miles down the coast. That day and the following day passed for us in a sort of nightmare. Our mouths were dry and our tongues were swollen. The wind was still strong and the heavy sea forced us to navigate carefully, but any thought of our peril from the waves was buried beneath the consciousness of our raging thirst. The bright moments were those when we each received our one mug of hot milk during the long, bitter watches of the night. Things were bad for us in those days, but the end was coming. The morning of May 8th broke thick and stormy, with squalls from the north-west. We searched the waters ahead for a sign of land, and though we could see nothing more than had met our eyes for many days, we were cheered by a sense that the goal was near at hand. About ten o'clock that morning we passed a little bit of kelp, a glad signal of the proximity of land. An hour later we saw two shags sitting on a big mass of kelp, and knew then that we must be within ten or fifteen miles of the shore. These birds are as sure an indication of the proximity of land as a lighthouse is, for they never venture far to sea. We gazed ahead with increasing eagerness, and at 12.30 p.m., through a rift in the clouds, McCarthy caught a glimpse of the black cliffs of South Georgia, just fourteen days after our departure from Elephant Island. It was a glad moment. Thirst-ridden, chilled, and weak as we were, happiness irradiated us. The job was nearly done.

We stood in towards the shore to look for a landing-place, and presently we could see the green tussock-grass on the ledges above the surf-beaten rocks. Ahead of us and to the south, blind rollers showed the presence of uncharted reefs along the coast. Here and there the hungry rocks were close to the surface, and over them the great waves broke, swirling viciously and spouting thirty and forty feet into the air. The rocky coast appeared to descend sheer to the sea. Our need of water and rest was wellnigh desperate, but to have attempted a landing at that time would have been suicidal. Night was drawing near, and the weather indications were not favourable. There was nothing for it but to haul off till the following morning, so we stood away on the starboard tack until we had made what appeared to be a safe offing. Then we hove-to in the high westerly swell. The hours passed slowly as we waited the dawn, which would herald, we fondly hoped, the last stage of our journey. Our thirst was a torment and we could scarcely touch our food; the cold seemed to strike right through our weakened bodies. At 5 a.m. the wind shifted to the north-west and quickly increased to one of the worst hurricanes any of us had ever experienced. A great cross-sea was running, and the wind simply shrieked as it tore the tops off the waves and converted the whole seascape into a haze of driving spray. Down into valleys, up to tossing heights, straining until her seams opened, swung our little boat, brave still but labouring heavily. We knew that the wind and set of the sea was driving us ashore, but we could do nothing. The dawn showed us a storm-torn ocean, and the morning passed without bringing us a sight of the land; but at 1 p.m., through a rift in the flying mists, we got a glimpse of the huge crags of the island and realized that our position had become desperate. We were on a dead lee shore, and we could gauge our approach to the unseen cliffs by the roar of the breakers against the sheer walls of rock. I ordered the double-reefed mainsail to be set in the hope that we might claw off, and this attempt increased the strain upon the boat. The *James Caird* was bumping heavily, and the water was pouring in everywhere. Our thirst was forgotten in the realization of our imminent danger, as we bailed unceasingly, and adjusted our weights from time to time; occasional glimpses showed that the shore was nearer. I knew that Annewkow Island lay to the south of us, but our small and badly marked chart showed uncertain reefs in the passage between the island and the mainland, and I dared not trust it, though as a last resort we could try to lie under the lee of the island. The afternoon wore away as we edged down the coast, with the thunder of the breakers in our ears. The approach of evening found us still some distance from Annewkow Island, and, dimly in the twilight, we could see a snow-capped mountain looming above us. The chance of surviving the night, with the driving gale and the implacable sea forcing us on to the lee shore, seemed small. I think most of us had a feeling that the end was very near. Just after 6 p.m., in the dark, as the boat was in the yeasty backwash from the seas flung from this iron-bound coast, then, just when things looked their worst, they changed for the best. I have marvelled often at the thin line that divides success from failure and the sudden turn that leads from apparently certain disaster to comparative safety. The wind suddenly shifted, and we were free once more to make an offing. Almost as soon as the gale eased, the pin that locked the mast to the thwart fell out. It must have been on the point of doing this throughout the hurricane, and if it had gone nothing could have saved us; the mast would have snapped like a carrot. Our backstays had carried away once before when iced up, and were not too strongly fastened now. We were thankful indeed for the mercy that had held that pin in its place throughout the hurricane.

We stood offshore again, tired almost to the point of apathy. Our water had long been finished. The last was about a pint of hairy liquid, which we strained through a bit of gauze from the medicine chest. The pangs of thirst attacked us with redoubled intensity, and I felt that we must make a landing on the following day at almost any hazard. The night wore on. We were very tired. We longed for day. When at last the dawn came on the morning of May 10th there was practically no wind, but a high cross-sea was running. We made slow progress towards the shore. About 8 p.m. the wind backed to the north-west and threatened another blow. We had sighted in the meantime a big indentation which I thought must be King Haakon Bay, and I decided that we must land there. We set the bows of the boat towards the bay and ran before the freshening gale. Soon we had angry reefs on either side. Great glaciers came down to the sea and offered no landing-place. The sea spouted on the reefs and thundered against the shore. About noon we sighted a line of jagged reef, like blackened teeth, that seemed to bar the entrance to the bay. Inside, comparatively smooth water stretched eight or nine miles to the head of the bay. A gap in the reef appeared, and we made for it. But the fates had another rebuff for us. The wind shifted and blew from the east right out of the bay. We could see the way through the reef, but we could not approach it directly. That afternoon we bore up, tacking five times in the strong wind. The last tack enabled us to get through, and at last we were in the wide mouth of the bay. Dusk was approaching. A small cove, with a boulder-strewn beach guarded by a reef, made a break in the cliffs on the south side of the bay, and we turned in that direction. I stood in the bows directing the steering as we ran through the kelp and made the passage of the reef. The entrance was so narrow that we had to take in the oars, and the swell was piling itself right over the reef into the cove; but in a minute or two we were inside, and in the gathering darkness the *James Caird* ran in on a swell and touched the beach. I sprang ashore with the short painter and held on when the boat went out with the backward surge. When the *James Caird* came in again, three of the men got ashore, and they held the painter while I climbed some rocks with another line. A slip on the wet rocks twenty feet up nearly closed my part of the story just at the moment when we were achieving safety. A jagged piece of rock held me and at the same time bruised me sorely. However, I made fast the line, and in a few minutes we were all safe on the beach with the boat

floating in the surging water just off the shore. We heard a
gurgling sound that was sweet music in our ears, and, peering
around, found a stream of fresh water almost at our feet. A
moment later we were down on our knees drinking the pure ice-
cold water in long draughts that put new life into us.

13
Anson in the Pacific

Commodore Anson sailed round the world between 1740 and 1744, with a squadron of ships. His main purpose was to capture the Spanish treasure ships homeward bound from South America. This was the age of piracy on the high seas, and the author, who was the chaplain on the *Centurion*, seems not to be worried on this account.

I have included this story because it shows how unwieldy the ships were in those days. This was well shown by the account of a ship sighting land, but then spending a whole month trying to beat up to it. During this time they were short of provisions and water, with members of the crew dying daily from scurvy, while help could come only from the shore which they could see but could not reach.

A. R.

from *ANSON'S VOYAGE ROUND THE WORLD*
by RICHARD WALTER

BEING arrived, on the 8th of May, off the Island of Socoro, which was the first rendezvous appointed for the squadron, and where we hoped to have met with some of our companions, we cruised for them in that station several days. But here we were not only disappointed in our expectations of being joined by our friends, and were thereby induced to favour the gloomy suggestions of their having all perished; but we were likewise perpetually alarmed with the fears of being driven on shore upon this coast which appeared too craggy and irregular to give us the least prospect, that in such a case any of us could possibly escape immediate destruction. For the land had indeed a most tremendous aspect: The most distant part of it, and which appeared far within the country, being the mountains usually called the Andes or Cordilleras, was extremely high, and

covered with snow; and the coast itself seemed quite rocky and
barren, and the water's edge skirted with precipices. In some
places indeed we discerned several deep bays running into the
land, but the entrance into them was generally blocked up by
numbers of little islands; and thought it was not improbable but
there might be convenient shelter in some of those bays, and
proper channels leading thereto; yet, as we were utterly ignorant of the coast, had we been driven ashore by the western
winds which blew almost constantly there, we did not expect to
have avoided the loss of our ship, and of our lives.

This continued peril, which lasted for above a fortnight, was
greatly aggravated by the difficulties we found in working the
ship; as the scurvy had by this time destroyed so great a part of
our hands, and had in some degree affected almost the whole
crew. Nor did we, as we hoped, find the winds less violent, as we
advanced to the northward; for we had often prodigious squalls
which split our sails, greatly damaged our rigging, and endangered our masts. Indeed, during the greatest part of the
time we were up on this coast, the wind blew so hard, that in
another situation, where we had sufficient sea-room, we should
certainly have lain-to; but in the present exigency, we were
necessitated to carry both our courses and topsails, in order to
keep clear of this lee-shore. In one of these squalls, which was
attended by several violent claps of thunder, a sudden flash of
fire darted along our decks, which, dividing, exploded with a
report like that of several pistols, and wounded many of our
men and officers as it passed, marking them in different parts of
the body: This flame was attended with a strong sulphureous
stench, and was doubtless of the same nature with the larger
and more violent blasts of lightning which then filled the air.

It were endless to recite minutely the various disasters,
fatigues, and terrors, which we encountered on this coast; all
these went on increasing till the 22nd of May, at which time, the
fury of all the storms which we had hitherto encountered,
seemed to be combined, and to have conspired our destruction.
In this hurricane almost all our sails were split, and great part of

our standing rigging broken; and, about eight in the evening, a mountainous overgrown sea took us upon our starboard quarter, and gave us so prodigious a shock, that several of our shrouds broke with the jerk, by which our masts were greatly endangered; our ballast and stores too were so strangely shifted, that the ship heeled afterwards two streaks to port. Indeed it was a most tremendous blow, and we were thrown into the utmost consternation from the apprehension of instantly foundering; and though the wind abated in a few hours, yet, as we had no more sails left in a condition to bend to our yards, the ship laboured very much in a hollow sea, rolling gunwale to, for want of sail to steady her: So that we expected our masts, which were now very slenderly supported, to come by the board every moment. However, we exerted ourselves the best we could to stirrup out shrouds, to reeve new landyards, and to mend our sails; but while these necessary operations were carrying on, we ran great risque of being driven on shore on the island of Chloe, which was not far distant from us; but in the midst of our peril the wind happily shifted to the southward, and we steered off the land with the mainsail only, the Master and myself undertaking the management of the helm, while everyone else on board was busied in securing the masts, and bending the sails as fast as they could be repaired.

This was the last effort of that stormy climate; for in a day or two after we got clear of the land, and found the weather more moderate than we had yet experienced since our passing Streights Le Maire.

And now having cruised in vain for more than a fortnight in quest of the other ships of the squadron, it was resolved to take the advantage of the present favourable season, and the offing we had made from this terrible coast, and to make the best of our way for the Island of Juan Fernandes. For though our next rendezvous was appointed off the harbour of Baldivia, yet as we had hitherto seen none of our companions at this first rendezvous, it was not to be supposed that any of them would be found at the second: Indeed we had the greatest reason to suspect, that all but ourselves had perished. Besides, we were by this time reduced to so low a condition, that instead of attempting to attack the places of the enemy, our utmost hopes could only suggest to us the possibility of saving the ship, and some part of the remaining enfeebled crew, by our speedy arrival at Juan Fernandes; for this was the only road in that part of the world where there was any probability of our recovering our sick, or refitting our vessel, and consequently our getting thither was the only chance we had left to avoid perishing at sea.

Our deplorable situation then allowing no room for deliberation, we stood for the Island of Juan Fernandes; and to save time, which was now extremely precious (our men dying four, five, and six in a day), and likewise to avoid being engaged again with a lee-shore, we resolved, if possible, to hit the island upon a meridian. And, on the 28th of May, being nearly in the parallel upon which it is laid down, we had great expectations of seeing it: But not finding it in the position in which the charts had taught us to expect it, we began to fear that we had gone too far to the westward; and therefore, though the Commodore himself was strongly persuaded that he saw it on the morning of the 28th, yet his officers believing it to be only a cloud, to which opinion the haziness of the weather gave some kind of countenance, it was, on a consultation, resolved to stand to the eastward, in the parallel of the island; as it was certain, that by this course we should either fall in with the island, if we were already to the westward of it; or should at least make the mainland of Chili, from whence we might take a new departure, and assure ourselves, by running to the westward afterwards, of not missing the island a second time.

On the 30th of May we had a view of the Continent of Chili, distant about twelve or thirteen leagues; the land made exceeding high, and uneven, and appeared quite white; what we saw being doubtless a part of the Cordilleras, which are always covered with snow. Though by this view of the land we ascertained our position, yet it gave us great uneasiness to find that we had so needlessly altered our course, when we were, in all probability, just upon the point of making the island; for the mortality amongst us was now increased to a most dreadful degree, and those who remained alive were utterly dispirited by this new disappointment, and the prospect of their longer continuance at sea. Our water too began to grow scarce; so that a general dejection prevailed amongst us, which added much to the virulence of the disease, and destroyed numbers of our best men; and to all these calamities there was added this vexatious circumstance, that when, after having got a sight of the main, we tacked and stood to the westward in quest of the island, we were so much delayed by calms and contrary winds, that it cost us nine days to regain the westing, which, when we stood to the eastward, we ran down in two. In this desponding condition, with a crazy ship, a great scarcity of fresh water, and a crew so universally diseased, there were not above ten foremast men in a watch capable of doing duty, and even some of these lame, and unable to go aloft: Under these disheartening circumstances, we stood to the westward; and, on the 9th of June, at daybreak, we at last discovered the long-wished-for Island of Juan Fernandes, bearing N. by E. $\frac{1}{2}$ E., at eleven or twelve leagues distance. And though, on this first view, it appeared to be a very mountainous place, extremely ragged and irregular; yet as it was land, and the land we sought for, it was to us a most agreeable sight. Because at this place only could we hope to put a period to those terrible calamities we had so long struggled with, which had already swept away above half our crew, and which, had we continued a few days longer at sea, would inevitably have completed our destruction. For we were by this time reduced to so helpless a condition, that out of two hundred and odd men which remained alive, we could not, taking all our watches together, muster hands enough to work the ship on an emergency, though we included the officers, their servants, and the boys.

The wind being northerly when we first made the island, we kept plying all that day, and the next night, in order to get in with the land; and wearing the ship in the middle watch, we

had a melancholy instance of the almost incredible debility of
our people; for the lieutenant could muster no more than two
quartermasters, and six foremast men capable of working; so
that without the assistance of the officers, servants, and the boys,
it might have proved impossible for us to have reached the
island, after we had got sight of it; and even with this assistance
they were two hours in trimming the sails: To so wretched a
condition was a sixty-gun ship reduced, which had passed
Streights Le Maire but three months before, with between four
and five hundred men, almost all of them in health and vigour.

However, on the 10th in the afternoon, we got under the lee
of the island, and kept ranging along it, at about two miles
distance, in order to look out for the proper anchorage, which
was described to be in a bay on the north side. Being now nearer
in with the shore, we could discover that the broken craggy
precipices, which had appeared so unpromising at a distance,
were far from barren, being in most places covered with woods,
and that between them there were every where interspersed the
finest vallies, clothed with a most beautiful verdure, and
watered with numerous streams and cascades, no valley, of any
extent, being unprovided of its proper rill. The water too, as we
afterwards found, was not inferior to any we had ever tasted,
and was constantly clear. The aspect of this country thus
diversified, would, at all times, have been extremely delightful;
but in our distressed situation, languishing as we were for the
land and its vegetable productions (an inclination constantly
attending every stage of the sea-scurvy), it is scarcely credible
with what eagerness and transport we viewed the shore, and
with how much impatience we longed for the greens and other
refreshments which were then in sight; and particularly the
water, for of this we had been confined to a very sparing allowance a considerable time, and had then but five ton remaining
on board. Those only who have endured a long series of thirst,
and who can readily recall the desire and agitation which the
ideas alone of springs and brooks have at that time raised in
them, can judge of the emotion with which we eyed a large

cascade of the most transparent water, which poured itself from a rock near a hundred feet high into the sea, at a small distance from the ship. Even those amongst the diseased who were not in the very last stages of the distemper, though they had been long confined to their hammocks, exerted the small remains of strength that were left them, and crawled up to the deck to feast themselves with this reviving prospect. Thus we coasted the shore, fully employed in the contemplation of this enchanting landskip, which still improved upon us the farther we advanced. But at last the night closed upon us, before we had satisfied ourselves which was the proper bay to anchor in; and therefore we resolved to keep in soundings all night (we having then from sixty-four to seventy fathom), and to send our boat next morning to discover the road. However, the current shifted in the night, and set us so near the land that we were obliged to let go the best bower in fifty-six fathom, not half a mile from the shore.

At four in the morning, the cutter was dispatched with our third lieutenant to find out the bay we were in search of, who returned again at noon with the boat laden with seals and grass; for though the island abounded with better vegetables, yet the boat's crew, in their short stay, had not met with them; and they well knew, that even grass would prove a dainty, as indeed it was all soon and eagerly devoured. The seals too were considered as fresh provision, but as yet were not much admired, though they grew afterwards into more repute. For what rendered them less valuable at this juncture, was the prodigious quantity of excellent fish, which the people on board had taken, during the absence of the boat.

The cutter, in this expedition, had discovered the bay where we intended to anchor, which we found was to the westward of our present station; and, the next morning, the weather proving favourable, we endeavoured to weigh, in order to proceed thither; but though, on this occasion, we mustered all the strength we could, obliging even the sick, who were scarce able to keep on their legs, to assist us; yet the capstan was so weakly

manned, that it was near four hours before we hove the cable
right up and down. After which, with our utmost efforts, and
with many surges and some purchases we made use of to increase
our power, we found ourselves incapable of starting the anchor
from the ground. However, at noon, as a fresh gale blew towards the bay, we were induced to set the sails, which fortunately tripped the anchor; and then we steered along shore,
till we came abreast of the point that forms the eastern part of the
bay. On the opening of the bay, the wind that had befriended us
thus far, shifted, and blew from thence in squalls; but by means
of the head way we had got, we loofed close in, till the anchor
brought us up in fifty-six fathom.

Soon after we had thus got to our new berth, we discovered a
sail, which we made no doubt was one of our squadron; and on
its nearer approach we found it to be the *Tryal* sloop. We immediately sent some of our hands on board her, by whose assistance
she was brought to an anchor between us and the land. We
soon found that the sloop had not been exempted from the
same calamities which we had so severely felt; for her Commander, Captain Saunders, waiting on the Commodore, informed him, that out of his small complement, he had buried
thirty-four of his men; and those that remained were so universally afflicted with the scurvy, that only himself, his lieutenant,
and three of his men, were able to stand by the sails. The *Tryal*
came to an anchor within us, on the 12th, about noon, and we
carried our hawsers on board her, in order to moor ourselves
nearer in shore; but the wind coming off the land in violent
gusts, prevented our mooring in the berth we intended.

Indeed, our principal attention was employed in business
rather of more importance: For we were now extremely
occupied in sending on shore materials to raise tents for the
reception of the sick, who died apace on board, and doubtless
the distemper was considerably augmented, by the stench and
filthiness in which they lay; for the number of the diseased was
so great, and so few could be spared from the necessary duty of
the sails to look after them, that it was impossible to avoid a

great relaxation in the article of cleanliness, which had rendered the ship extremely loathsome between decks. Notwithstanding our desire of freeing the sick from their hateful situation, and their own extreme impatience to get on shore, we had not hands enough to prepare the tents for their reception before the 16th; but on that and the two following days we sent them all on shore, amounting to a hundred and sixty-seven persons, besides twelve or fourteen who died in the boats, on their being exposed to the fresh air. The greatest part of our sick were so infirm, that we were obliged to carry them out of the ship in their hammocks, and to convey them afterwards in the same manner from the water side to their tents, over a stony beach. This was a work of considerable fatigue to the few who were healthy; and therefore the Commodore, according to his accustomed humanity, not only assisted herein with his own labour, but obliged his officers, without distinction, to give their helping hand. The extreme weakness of our sick may in some measure be collected from the numbers who died after they had got on shore; for it had generally been found, that the land, and the refreshments it produces, very soon recover most stages of the sea-scurvy; and we flattered ourselves, that those who had not perished on this first exposure to the open air, but had lived to be placed in their tents, would have been speedily restored to their health and vigour. Yet, to our great mortification, it was near twenty days after their landing, before the mortality was tolerably ceased; and for the first ten or twelve days, we buried rarely less than six each day, and many of those who survived, recovered by very slow and insensible degrees. Indeed, those who were well enough, at their first getting on shore, to creep out of their tents, and crawl about, were soon relieved, and recovered their health and strength in a very short time; but, in the rest, the disease seemed to have acquired a degree of inveteracy, which was altogether without example.

The arrival of the *Tryal* sloop at this island so soon after we came there ourselves, gave us great hopes of being speedily joined by the rest of our squadron; and we were for some days continually looking out, in expectation of their coming in sight. But near a fortnight being elapsed, without any of them having appeared, we began to despair of ever meeting them again; as we knew, that had our ship continued so much longer at sea, we should every man of us have perished, and the vessel, occupied by dead bodies only, would have been left to the caprice of the winds and waves. And this we had great reason to fear was the fate of our consorts, as each hour added to the probability of these desponding suggestions.

But on the 21st of June, some of our people, from an eminence on shore, discerned a ship to leeward, with her courses even with the horizon; and they, at the same time, particularly observed that she had no sail abroad except her courses and her main topsail. This circumstance made them conclude that it was one of our squadron, which had probably suffered in her sails and rigging as severely as we had done: But they were prevented from forming more definite conjectures about her; for, after viewing her for a short time, the weather grew thick and hazy, and they lost sight of her. On this report, and no ship appearing for some days, we were all under the greatest concern, suspecting that her people were in the utmost distress for want of water, and so diminished and weakened by sickness, as not to be able to ply up to windward; so that we feared that, after having been in sight of the island, her whole crew would notwithstanding perish at sea. However, on the 26th towards noon, we discerned a sail in the north-east quarter, which we conceived to be the very same ship that had been seen before, and our conjectures proved true: And about one o'clock she approached so near, that we could distinguish her to be the *Gloucester*. As we had no doubt of her being in great distress, the Commodore immediately ordered his boat to her assistance, laden with fresh water, fish, and vegetables, which was a very seasonable relief to them; for our apprehensions of their calamities appeared to be but too well grounded, as perhaps there never was a crew in a more distressed situation. They had already thrown overboard two-thirds of their complement, and of those which remained alive, scarcely any were capable of doing duty except the officers and their servants. They had been a considerable time at the small allowance of a pint of fresh water to each man for twenty-four hours, and yet they had so little left, that, had it not been for the supply we sent them, they must soon have died of thirst.

The ship plied in within three miles of the bay; the winds and currents being contrary, she could not reach the road. However, she continued in the offing the next day; but as she had no chance of coming to an anchor, unless the winds and currents shifted, the Commodore repeated his assistance, sending to her the *Tryal*'s boat manned with the *Centurian*'s people, and a further supply of water and other refreshments. Captain Mitchel, the Captain of the *Gloucester*, was under the necessity of detaining both this boat and that sent the preceding day; for without the help of their crews, he had no longer strength enough to navigate the ship.

In this tantalizing situation the *Gloucester* continued for near a fortnight, without being able to fetch the road, though frequently attempting it, and at some times bidding very fair for it. On the 9th of July, we observed her stretching away to the eastward at a considerable distance, which we supposed was with a design to get to the southward of the Island but as we soon lost sight of her, and she did not appear for near a week, we were prodigiously concerned, knowing that she must be again in extreme distress for want of water. After great impatience about her, we discovered her again on the 16th, endeavouring to come round the eastern point of the Island: but the wind, still blowing directly from the bay, prevented her getting nearer than within four leagues of the land. On this, Captain Mitchel made signals of distress, and our long-boat was sent to him with a store of water and plenty of fish and other refreshments. And the long-boat being not to be spared, the Cockswain had positive orders from the Commodore to return again immediately; but the weather proving stormy the next day, and the boat not appearing, we much feared she was lost, which would have proved an irretrievable misfortune to us all: However, the third day after, we were relieved from this anxiety, by the joyful sight of the long-boat's sails upon the water; on which we sent the cutter immediately to her assistance, who towed her alongside in a few hours; when we found that the crew of our long-boat had taken in six of the *Gloucester*'s sick men to bring them on shore, two of which had died in the boat.

We now learnt that the *Gloucester* was in a most dreadful condition, having scarcely a man in health on board, except those they received from us: and, numbers of their sick dying daily, it appeared that, had it not been for the last supply sent by our long-boat, both the healthy and diseased must have all perished together for want of water. These calamities were the more terrifying, as they appeared to be without remedy: for the *Gloucester* had already spent a month in her endeavours to fetch the bay, and she was now no farther advanced than at the first moment she made the island; on the contrary, the people on board her had worn out all their hopes of ever succeeding in it, by the many experiments they had made of its difficulty. Indeed, the same day her situation grew more desperate than ever, for after she had received our last supply of refreshments, we again lost sight of her; so that we in general despaired of her ever coming to an anchor.

Thus was this unhappy vessel bandied about within a few leagues of her intended harbour, whilst the neighbourhood of that place and of those circumstances, which could alone put an end to the calamities they laboured under, served only to aggravate their distress, by torturing them with a view of the relief it was not in their power to reach. But she was at last delivered from this dreadful situation, at a time when we least expected it; for after having lost sight of her for several days, we were pleasingly surprised, on the morning of the 23rd of July, to see her open the NW. point of the bay with a flowing sail; when we immediately dispatched what boats we had to her assistance, and in an hour's time from our first perceiving her, she anchored safe within us in the bay.

I have thus given an account of the principal events, relating to the arrival of the *Gloucester*, in one continued narration. I shall only add, that we never were joined by any other of our ships, except our Victualler, the *Anna Pink*, who came in about the middle of August, and whose history I shall defer for the present; as it is now high time to return to the account of our own transactions on board and on shore, during the interval of the *Gloucester*'s frequent and ineffectual attempts to reach the island.

Our next employment, after sending our sick on shore from the *Centurian*, was cleansing our ship and filling our water. The first of these measures was indispensably necessary to our future health; as the numbers of sick, and the unavoidable negligence arising from our deplorable situation at sea, had rendered the decks most intolerably loathsome. And the filling our water was a caution that appeared not less essential to our security, as we had reason to apprehend that accidents might intervene, which would oblige us to quit the island at a very short warning; for some appearances we had discovered on shore upon our first landing, gave us grounds to believe, that there were Spanish cruisers in these seas, which had left the Island but a short time before our arrival, and might possibly return thither again, either for a recruit of water, or in search of us; since we could not doubt, but that the sole business they had at sea was to intercept us, and we knew that this island was the likeliest place, in their own opinion, to meet with us. The circumstances which gave rise to these reflections were, our finding on shore several pieces of earthen jars, made use of in those seas for water and other liquids, which appeared to be fresh broken: We saw, too, many heaps of ashes, and near them fish-bones and pieces of fish, besides whole fish scattered here and there, which plainly appeared to have been but a short time out of the water, as they were but just beginning to decay.

However, our fears on this head proved imaginary, and we were not exposed to the disgrace, which might have been expected to have befallen us, had we been necessitated (as we must have been, had the enemy appeared) to fight our sixty-gun ship with no more than thirty hands.

Whilst the cleaning of our ship and the filling our water went on, we set up a large copper oven on shore near the sick tents, in which we baked bread every day for the ship's company; for being extremely desirous of recovering our sick as soon as possible, we conceived that new bread, added to their greens and fresh fish, might prove a powerful article in their relief. Indeed, we had all imaginable reason to endeavour at the augmenting our present strength, as every little accident which to a full crew would be insignificant, was extremely alarming in our present helpless situation: Of this, we had a troublesome instance on the 30th of June; for at five in the morning, we were astonished by a violent gust of wind directly offshore, which instantly parted our small bower cable about ten fathom from the ring of the anchor: the ship at once swung off to the best bower, which happily stood the violence of the jerk, and brought us up with two cables on end in eighty fathom. At this time we had not above a dozen seamen in the ship, and we were apprehensive, if the squall continued, that we should be driven to sea in this wretched condition. However, we sent the boat on shore, to bring off all who were capable of acting; and the wind, soon abating of its fury, gave us an opportunity of receiving the boat back again with a reinforcement. With this additional strength we immediately went to work, to heave in what remained of the cable, which we suspected had received some damage from the foulness of the ground before it parted; and agreeable to our conjecture, we found that seven fathom and a half of the outer end had been rubbed, and rendered unserviceable. In the afternoon, we bent the cable to the spare anchor, and got it over the ship's side; and the next morning, July 1st, being favoured with the wind in gentle breezes, we warped the ship in again, and let go the anchor in forty-one fathom; the eastermost point now bearing from us E. ½ S.; the westermost NW. by W.; and the bay as before, SSW.; a situation in which we remained secure for the future. However, we were much concerned for the loss of our anchor, and swept frequently for it, in hopes to have recovered it; but the buoy having sunk at the very instant that the cable parted, we were never able to find it.

And now as we advanced in July, some of our men being tolerably recovered, the strongest of them were put upon cutting down trees, and splitting them into billets: while others, who were too weak for this employ, undertook to carry the billets by one at a time to the water-side: This they performed, some of them with the help of crutches, and others supported by a single stick. We next sent the forge on shore, and employed our smiths, who were but just capable of working, in mending our chain-plates, and our other broken and decayed ironwork. We began too the repairs of our rigging; but as we had not junk enough to make spun-yarn, we deferred the general overhale, in hopes of the daily arrival of the *Gloucester*, who we knew had a great quantity of junk on board. However, that we might dispatch as fast as possible in our refitting, we set up a large tent on the beach for the sail-makers; and they were immediately employed in repairing our old sails, and making us new ones. These occupations, with our cleansing and watering the ship (which was by this time pretty well completed), the attendance on our sick, and the frequent relief sent to the *Gloucester*, were the principal transactions of our infirm crew, till the arrival of the *Gloucester* at an anchor in the bay.

And now, after the *Gloucester*'s arrival, we were employed in earnest in examining and repairing our rigging: but in the stripping our foremast, we were alarmed by discovering it was sprung just above the partners of the upper deck. The spring was two inches in depth, and twelve in circumference; however, the carpenters, on inspecting it, gave it as their opinion, that fishing it with two leaves of an anchor-stock, would render it as secure as ever. But, besides this defect in our mast, we had other difficulties in refitting, from the want of cordage and canvas; for though we had taken to sea much greater quantities of both, than had ever been done before, yet the continued bad weather we met with had occasioned such a consumption of these stores, that we were driven to great straits: As after working up all our junk and old shrouds, to make twice-laid cordage, we were at last obliged to unlay a cable to work into running rigging. And with all the canvas, and remnants of old sails that could be mustered, we could only make up one complete suit.

Towards the middle of August our men being indifferently recovered, they were permitted to quit their sick tents, and to build separate huts for themselves, as it was imagined, that by living apart, they would be much cleanlier, and consequently likely to recover their strength the sooner; but at the same time particular orders were given, that, on the firing of a gun from the ship, they would instantly repair to the water-side. Their employment on shore was now either the procuring of refreshments, the cutting of wood, or the making of oil from the blubber of the sea-lions. This oil served us for several purposes, as burning in lamps, or mixing with pitch to pay the ship's sides, or, when worked up with wood ashes, to supply the use of tallow (of which we had none left) to give the ship boot-hose tops. Some of the men too were occupied in salting of cod; for there being two *Newfoundland* fishermen in the *Centurian*, the Commodore set them about laying in a considerable quantity of salted cod for a sea-store, though very little of it was used, as it was afterwards thought to be as productive of the scurvy as any other kind of salt provisions.

I have before mentioned, that we had a copper oven on shore to bake bread for the sick; but it happened that the greatest part of the flour, for the use of the squadron, was embarked on board our victualler the *Anna Pink*: And I should have mentioned, that the *Tryal* sloop, at her arrival, had informed us, that on the 9th of May, she had fallen in with our victualler, not far distant from the Continent of Chili; and had kept company with her for four days, when they were parted in a hard gale of wind. This afforded us some room to hope that she was safe, and that she might join us; but all June and July being past, without any news of her, we then gave her over for lost, and at the end of July, the Commodore ordered all the ships to a short allowance of bread. Nor was it in our bread only that we feared a deficiency; for since our arrival at this Island, we discovered that our former purser had neglected to take on board large quantities of several kinds of provisions, which the Commodore had expressly ordered him to receive; so that the supposed loss of our victualler was on all acounts a mortifying consideration. However, on Sunday, the 16th of August, about noon, we espied a sail in the northern quarter, and a gun was immediately fired from the *Centurian* to call off the people from shore; who readily obeyed the summons, repairing to the beach, where the boats waited to carry them on board. And being now prepared for the reception of this ship in view, whether friend or enemy, we had various speculations about her; some judging it to be the *Severn*, others the *Pearl*, and several affirming that it did not belong to our squadron: But about three in the afternoon our disputes were ended, by an unanimous persuasion that it was our victualler the *Anna Pink*. This ship though, like the *Gloucester*, she had fallen in to the northward of the island, had yet the good fortune to come to an anchor in the bay, at five in the afternoon. Her arrival gave us all the sincerest joy; for each ship's company was immediately restored to their full allowance of bread, and we were now freed from the apprehensions of our provisions falling short, before we could reach some amicable port; a calamity which in these seas is of all others the most irretrievable.

14
Capsize in a Typhoon

At the time of this adventure (1911) Voss was sealing out of Yokohama, but a fifteen-year ban had recently been put on seal killing
and, while waiting for the compensation due to him from the Japanese Government, he elected to start on a voyage round the world
with two young men who had just taken delivery of a new yawl,
some 25 ft in overall length. Voss was an experienced small ship
sailor, having sailed almost round the world in a converted Red
Indian canoe. He had little fear of the sea, believing that in the worst
weather oil poured on the water and the use of his sea anchor would
bring any ship through the worst weather.

His almost light-hearted description of *Sea Queen* meeting with a
typhoon suggested to some that the author might be romancing a
little: but a well-known yachtsman met the *Sea Queen* on her arrival
at Yokohama and, after inspecting the damage, he never doubted
the truth of Voss's account of what she had been through.

A. R.

from '*THE VENTURESOME VOYAGES OF CAPTAIN VOSS*'
by JOHN CLAUS VOSS

THE *Sea Queen*, after completing repairs, and with everything ready for sea again, left Aikawa on August 22nd at
half-past three in the afternoon, and proceeded on her course
towards the Marshall Islands. When outside the harbour we
steered south-east by east. The weather was dull, with a heavy
swell setting from the north-east. The wind being light, our
progress was slow, and at midnight Kinkasan light was bearing
north by west, distant nineteen miles. On the following two days
we experienced light and variable winds accompanied by spells
of calm. August 25th the *Sea Queen*'s usual head wind arrived,

starting to blow from the east and accompanied by a choppy sea. Similar conditions were experienced on the 26th and 27th, the wind blowing from the east with increasing seas and heavy rain squalls. The boat was kept going to the south-east.

On August 28th our position at noon was latitude thirty-two degrees forty minutes north and longitude one forty-five degrees five minutes east. This day came in with a strong south-easterly breeze, accompanied by heavy rain squalls, and a very large swell set in from the same direction. Towards noon the weather cleared and the wind moderated, but the swell kept on increasing. According to the state of the weather and the swell setting in from the south-east I said to myself, 'I should not be at all surprised if we get a change, and that it will be from bad to worse,' as according to indications a typhoon was approaching. I, however, kept my thoughts to myself till the following day, August 29th, which was ushered in with a moderate easterly breeze and clear weather. Apparently Father Neptune was trying to make things a little easier for the crew of the *Sea Queen*, but the large south-easterly swell continued, and at about nine o'clock a large heavy-looking ring of varied fiery colours formed round the sun. The atmosphere became very sultry, and the appearance of a dense bank of clouds of threatening appearance on the horizon convinced me that the dread visitor was close at hand, and that I would be called upon to prove the statement I had so often made, i.e., that a small vessel is as safe as a large one.

It was then that I spoke to my two shipmates about the approaching typhoon. The large circle and the similar appearance of the sky followed the sun round till about four in the afternoon, after which the sun disappeared behind the cloud-bank on the horizon. Between sunset and dark the clouds which covered the sky became a fiery hue. The weather during the night was warm and pleasant, and anyone who does not understand the indications of a typhoon would not have thought that, inside of thirty hours, the *Sea Queen* would lay in the centre of one of the worst typhoons that has ever blown in the North Pacific. When darkness set in and the angry clouds had disappeared, the only indication of the approaching typhoon was the south-easterly swell and the barometer, which, however, was not a low glass by any means. At eight o'clock it stood at twenty-nine eighty, but was going down steadily, and between this and the increase in size of the tremendous swell, which almost began to break, I was sure that the approaching hurricane was not far off.

August 30th came in with a light, baffling wind, flying about between east and south, and accompanied by a dark, cloudy sky. At six o'clock the wind settled in the south-east with increasing force and occasional heavy rain squalls. At noon Port Lloyd of the Bonin group bore south by west-half-west two hundred and forty-five miles. An hour later, as the wind was still increasing and the squalls getting heavier, we laid the boat to under a reefed mizzen and storm stay-sail, but at two o'clock we were obliged to put the little vessel under sea anchor and riding sail. The barometer registered twenty-nine forty-five and was falling steadily.

During the afternoon and night the wind continued to increase in force, accompanied by heavy rain squalls and breaking seas, and by two o'clock the following morning the force of the wind was that of a heavy gale, and as the seas were breaking badly we hung two oil bags over the side, which we changed every hour. The proud little *Sea Queen* rode the large seas as they came along; now and then a sea would break near us, and she would give an extra little roll. At eight o'clock it was blowing so hard that I did not think it possible for the wind and sea to increase, and seeing that the boat was laying head to wind, with her drag ahead and a reefed mizzen over her stern, and in spite of the large seas lying quite comfortably, I told my shipmates that the little vessel would weather the typhoon without much trouble; and I am certain she would have done so had our storm gear been stronger.

After making the cruises in the *Xora* and *Tilikum*, I thought I surely knew all about the sea and what was required to manage a small vessel through a heavy gale. As far as heavy gales are concerned, I am quite sure I did understand the management of a boat, but was convinced that morning that I still had to learn something about typhoons. At nine the wind blew with such force that it was impossible to stand on deck, and we were obliged to lie flat down in the cockpit and hang on for dear life. Still the gallant little boat kept facing the gale, and it was most wonderful how she got over the top of the tremendous seas. Up to an hour previously the oil bags did quite a lot of good in keeping the break off the large seas, but now the oil seemed to have little effect, and no trace could be detected on the water. Nevertheless my two mates were employed in the cabin getting the bags ready, while I looked after them outside, thinking that if the oil did no good it would not be harmful. The water was then flying over the boat like a heavy snowdrift, and it was impossible to look up against it, but fortunately no heavy seas came aboard, and there was therefore no danger of the vessel foundering. Vincent and Stone every now and then would open the cabin scuttle a little to inquire how we were getting along. My reply was always the same – 'All right.'

Shortly after nine, when I noticed the boat fell sideways into the sea, the mizzen sheet parted. 'All hands on deck!' I yelled. In a second my two mates were alongside me. All three of us were crawling about the deck on all fours on the after-end of the vessel, and with the seas washing over us we managed to take in the mizzen and save the sail. We then found that our sea anchor was lost. Stone managed to get forward and pull the anchor rope on board, and Vincent and I got busy and made a temporary sea anchor by tying together the cabin ladder and one of our anchors. All this had to be done while lying flat upon deck. After fastening the anchor line to the temporary sea anchor we dropped it overboard, and as it was impossible to set the mizzen again we were obliged to let the vessel lay to this temporary arrangement alone. Owing to the fact that this sea anchor was not much of a drag, and being without the aid of the storm sail, it did not do much good. Shortly before eleven o'clock we lost the temporary sea anchor also, and the boat then lay sideways to the sea. The rudder up to this time had been lashed amidships, but as the boat was lying sideways to the sea, I took the rudder lashings off to allow it to swing about and give her a better change to drift sideways with the seas.

My two mates were down in the cabin again and I was lying in the cockpit holding on with one hand and with the other keeping the oil bags in the water when a huge sea struck the boat and put her on her beam ends, in which position she remained for just a second or two. I wondered what would happen next; whether she would recover or turn turtle. I was not left long in doubt, for I then felt a little jerk which told me that the boat was turning bottom up, and to save myself from getting under her I let go my hold. The next moment I was in the water, and felt certain that it was all up with us; in fact, I took two big mouthfuls of water, thinking to go down quickly and be done with it. When a man gets into a fix of that kind, however, many thoughts will run through his mind, and after I had said good-bye to the world and taken in the water ballast I thought of my two young shipmates who were inside the boat and unable to get out, and wished just to see them once more to say good-bye. By that time I had been under water long enough to be dead, but wasn't. I popped up again behind the stern of the boat and saw the *Sea Queen* in front of me, with her keel pointing to the sky. I made one kick and grabbed hold of her stern, and then made up my mind to get on her bottom and do something towards righting her again.

I have heard it said that 'While there's life there's hope,' and 'Where there's a will there's a way.' At that time I was still alive and had a little will left, but I thought it was going hard with the hopes and ways. Anyway, I made use of the little determination that remained and climbed over the stern of the boat. Just as I got on the bottom I saw an enormous breaking sea coming towards us, so I dug my nails into the keel to avoid getting washed off again. In a second the sea struck the boat along the keel, but I managed to hang on. The same sea caused the boat to heel over, and the weight of the iron on the keel slowly but surely brought her right side up again, and as she turned over I scrambled over the gunwale, and by the time the little vessel was floating on her bottom again I was in the cockpit. The next thing I saw the scuttle open, and heard Vincent shouting at the top of his voice, 'Are you there, Captain?' And then my two shipmates leaped out of the cabin.

No doubt some of my readers have seen two porpoises jumping out of the water at sea, one after the other. Well, it looked just like that to me. In spite of being on a small craft which was broadside on to the worst storm and largest sea that I ever experienced in all my years at sea, and all three of us lying down in the cockpit and hanging on for dear life, our meeting after the incident was quite joyous, and if the boat got smashed up in the typhoon, it would have given us a chance to say good-bye to each other.

The storm was then at its worst, and the force of the wind was that of several concentrated heavy gales. Large seas were washing over us in rapid succession. Owing to the wind, the heavy rain and salt-water spray, we were absolutely unable to open our eyes when facing windward, and could only see about a hundred feet to leeward. Every few minutes the little vessel was heeled over with the mast in the water. The vessel and gear up to that time was still in good order, and nothing had been lost or broken, but in turning upside-down a tremendous amount of water got into her. We were lying with our starboard side to the sea and wind, and as our cabin scuttle was on the port side it was taken under water every time the boat was thrown on her beam ends, and consequently much water got into the cabin. It was impossible to open the cabin door while the boat was lying on the starboard tack, and we were therefore unable to bale her out. At the same time I was certain that if she kept in the same position for another twenty minutes she would have filled with water and foundered; so the only thing that could possibly be done was to put the boat on the port tack, and owing to the condition we were in it had to be done quickly.

Consequently, I told my mates to hold on tight so that none of us would get washed overboard during the operation. I then put the helm up to get headway on the boat, and to make her steer I slacked the mainboom off. No sooner was that done than the little vessel started to go ahead, and in a few seconds we were round on the other tack. Everything had gone wonderfully well until the boat came sideways to the sea, and as she was still going ahead a sea broke over the top of us and smashed the mainboom. This was the first thing, apart from the sea anchor, that had got broken until then.

The next thing to be done was to bale some of the water out of the boat. Stone and Vincent both knew what it was like to be in the cabin when she turned bottom up, and both being good swimmers elected to take their chance on deck. I myself, being a very poor swimmer, stood a better chance inside; so it was therefore agreed that Vincent and Stone would tend the scuttle, keeping it open when there was a chance, and I went below to bale her out. It was also agreed that in the event of the boat turning bottom up again my two faithful, athletic shipmates would get on the lee side of her bottom and help the large seas to right her again.

The scuttle of the cabin door was then opened and I made a dive into the cabin. No sooner was I inside than the scuttle was closed over my head and the boat thrown on her beam. I felt sure that she would turn bottom up again, and with all the water that was already inside I thought it must be the last of us. To my surprise, however, she righted herself to an angle of about forty-five degrees, which of course raised the scuttle out of the water, and it was at once slightly opened by my mates on deck. I shall never forget the sight that met my eyes in the cabin. Apart from having a good opinion of my ability to handle boats at sea, I also thought I had learnt all there was to know about stowing stores and cargo so that they could not shift in any weather. In this I was rudely disillusioned, and consequently I came to the conclusion that I was adding to my experiences during that typhoon in more than one sense.

The boat was just about one-third full of water, in which were floating pieces of our gramophone and many records, amongst them 'Life on the Ocean Wave'. Our photographic outfit, tinned goods of various descriptions, bedding, blankets, books, clothing, silver watches, gold chains and navigation instruments were washing about in a mixture of fish oil, kept for use against stormy seas, which had got upset amongst the lot. Now the smell of fish oil is about the least dainty perfume I can possibly think of. The smell and the rolling of the boat, with everything washing from one side of the cabin to the other, and myself amongst it, almost turned me sea-sick. But I gave my stomach to understand that I could not afford to be particular just then, but was obliged to set to work to get some of the water out. For a baler I used a five-pound sugar tin, and whenever my mates outside had an opportunity to open the scuttle far enough I threw some water out. I was certainly baling under difficulties, for nearly every minute the boat was thrown on her beam ends and the scuttle had to be closed till she righted again; then, instantly, the scuttle would be opened, and I emptied the tin as fast as I possibly could.

We had kept up the performance for about an hour, and as I had made very little impression on the water in the cabin I thought that the little vessel must have sprung a leak, but as it was a matter of life or death we kept at it. A little later Vincent opened the scuttle just enough to shout through the opening, 'Stone is overboard!' and then quickly closed it to keep the water from pouring into the cabin, for again the boat was put on her beam ends. Next time she righted the scuttle was opened and Vincent shouted, 'He is on board again.' Of course I was quite certain that neither of my shipmates would get away from the boat as long as she kept afloat.

We had been working hard for three hours, and in spite of the boat being thrown on her beam ends time after time she never turned turtle again, and the water was then pretty well out of her. Vincent opened the scuttle again and said, 'Captain, I think we are in the centre of the typhoon.' 'No, not in the centre yet,' I answered, 'but I think we will be shortly, and the sooner we get there the better.' I knew by that time that in spite of the tremendous size of the breaking seas they could not harm our little vessel, but that it was the terrible strength of the wind that forced her bottom upwards, and put her on her beam ends time after time. I knew also that in the centre of the typhoon there would be no wind, and we should be all right.

Stone got washed overboard a second time, and the nerve-racking experiences were now beginning to tell upon both my companions, so, as the cabin was free of water and nothing further could be done on deck, I advised them to come down into the cabin. We then shut the door up as tightly as we could and waited for further developments.

Our timepieces being out of action, we did not know the exact hour, but I think it was between two and three in the afternoon that the barometer registered twenty-eight twenty-five, and after the three of us had been sitting down some time expecting the boat to turn turtle again any second, we were surprised at her righting herself on an even keel. I at once opened the scuttle, and found that we were in a dead calm, and that both masts and all the gear were overboard. All three of us lost no time in getting the broken spars, gear and sails on board again. We were then in the centre of the typhoon.

I have heard ship captains say that once a vessel gets into the centre of the typhoon she will never get out again. After my experience, I can quite agree where a large vessel is concerned, as with such tremendous seas and without a breath of wind she would roll herself to pieces. But the *Sea Queen* did not mind it in the least; she just simply bobbed up and down as the seas came along. I never had a better chance to prove my statement, both while the typhoon was blowing and after we got into the centre of it, that in a heavy gale, be they large or small, as long as their headway is stopped and they are allowed to take a natural drift with the sea and wind, a sea has no power on vessels, and if it does roll on board a large vessel there is no force behind it.

After we got all our gear aboard and well secured on deck we cleared up the cabin; and of course everything was soaked with salt water and fish oil. You may talk of the odour of bad eggs; it isn't in it with the smell of fish oil. All our bedding was saturated, and consequently we were obliged to transfer it to the deck and manage without for the time being. We only had one thing dry in the boat: that was our matches, which we kept in sealed bottles. Our stove was similar to the one I used on the *Tilikum*, which, like my sea anchor, never failed me. On this occasion, however, the sea anchor had failed to do its duty, but the stove, in spite of being tumbled about in the cabin, was there ready when wanted, and in a very short time we had water boiling and made a good cup of coffee, which put us all three in good humour again. This was the first time in all my experiences in small vessels at sea that I was obliged, owing to the weather, to go without my regular meals.

As the barometer remained stationary, we were certain that the second half of the typhoon was near at hand. We were not mistaken, for in about half an hour's time we entered the second stage. It then blew very hard for about four hours from the west, but with the spars out of her it was a picnic for the little vessel. Shortly after the wind changed the barometer started to rise, and by midnight showed twenty-eight fifty-five. During the following twenty-four hours the weather gradually cleared, but as there was still a large sea and strong wind from the west we were unable to do anything towards repairing the damage.

September 2nd came in with a moderate breeze and a large westerly swell, and as the boat was still tumbling about quite a lot we were unable to do anything to our main-mast, but managed to lash together a piece of the broken mizzen-mast and spinnaker-boom, and after some difficulty we managed to step the same in place of the main-mast. Owing to the flimsiness of our temporary jury-mast, all it would carry was the storm stay-sail. During that day and night we experienced fresh south-west breezes, and with our staysail made a mile and a half an hour to the north-north-west. The distance from Yokohama was about three hundred and fifty miles.

September 3rd came in with a moderate south-westerly breeze. At daybreak the wind became very light, and as the sea was quite smooth we went to work brightly and early to look over the damage that had been done, and to repair it as best we could. In our survey we found that the main-mast had been broken just above the deck and about three feet below the eyes of the rigging, the main-boom about four feet from the inner end, and the mizzen-mast, like the main-mast, just above the deck, about five feet higher up and again just below the eyes of the rigging. The rigging and chain-plates were in good order, so the two masts had simply buckled up and by the force of the wind blown out of the boat in pieces. The rudder post was broken in the top part of the rudder blade. The hull of the boat was as good as ever.

The dimensions of the main-mast were: Length from deck to eyes of shrouds, twenty-one feet nine inches. Diameter at the lower break, five inches. Diameter at upper break, four and a half inches. Length of mizzen-mast from deck to eyes of shrouds, fifteen feet. Diameter at lower break, four inches; at next break, four inches; and at the eyes of the rigging, three inches.

Both masts were made of Japanese pine, and at the time they blew out of the boat she was lying sideways to the sea and wind at an angle of about forty-five degrees away from the wind. Now, I should like to know what was the speed of the wind when the masts were broken!

Having cleared the rigging, ropes, sails, etc., from the deck, we laid the two broken ends of the foremast together, after which we took half-inch pine boards, which had served until then as inside planking along the side of the boat. These boards we cut into narrow strips three feet long and nailed them side by side around the mast. We then put five lashings over the nailed-on strips of wood at an equal distance apart. This is called 'fishing a spar' and if properly done the spar will be as strong as ever, which was the case with us. Owing to the lower break being near the deck, it could not be 'fished' very well without enlarging the mast hole through the deck, and as we did not wish to do that, we cut a step on the mast at the lower break, which of course made it that much shorter. It was then about eleven o'clock and the main-mast was ready to be stepped again.

Ever since the typhoon the talk in the *Sea Queen* had been in connection with getting the main-mast into the vessel again, and as we had such difficulty in stepping the first jury-mast, my two young mates had almost given it up for a bad job, and when I repeatedly told them that to men like us, with good hands, it was nothing, they replied that I was blowing my whistle again, as I had done about eating raw fish.

The main-mast was lying on the deck with all rigging, blocks and gear attached, ready to be stepped, and we proceeded as follows. We took the small jury-mast down, and used the spinnaker-boom for one sheerleg and part of the main-boom for the other. We then put on the sheer lashing, fore- and aft-guys and a tackle to raise the mast with. The sheerlegs being in place, we set up the fore-guy to the end of the bowsprit, and the after-guy to the end of the boom-kin. The lower ends of the sheerlegs we lashed solidly to the chain-plates. The sheerlegs were thus so secured that not even a typhoon could have blown them overboard. The lower block of the tackle was then fastened about eight feet from the lower end of the mast, and when everything was ready to hoist, and as things then looked very prosperous for the *Sea Queen* and her crew to get back to mother earth, I said to the mate, 'By the way, Mr Stone, just before we put in this mast we must have a little understanding about the promise you made the other day. You remember when we were in the centre of the typhoon and just after the masts blew out, you said something about if we got safely back to land you would treat the crew to a champagne dinner. Now I would like to know if that promise is still good.'

'I shall certainly keep my word,' he replied.

'Mr Vincent,' I said, 'you heard that?'

'I did,' said Vincent, 'and shall see that he carries it out.'

'Hoist away on the mast halliard,' I said, and up went the mast.

Owing to the sheerlegs being too short to sling the mast in her balance, we were obliged to fasten the tackle in such a position that the lower end of the spar would just clear the deck when it was hauled down to be stepped. Consequently, when the mast was hoisted up the top was lying on the after end of the deck, and the lower end was sticking up over the bow. Just then the vessel got broadside to the swell and started rolling about, which certainly made it very difficult to step the mast, so by means of our oars we brought the boat head to sea, and then, by pulling the lower end of the mast down by means of a rope it was stepped almost as quickly as I can write it. It was a great relief for the crew of the little boat to see a mast in her again. Of course, it was about three feet shorter than originally, and the splice below the eyes of the rigging gave it the appearance of a horse's leg in splints. Still, it was a mast; and we were well satisfied, since it was large and strong enough to take us back to land. Our medicine chest still contained some brandy, and we would have celebrated the occasion but for the fact that there was a lot of other work to be done before we were able to set sail again; so we had a little tiffin instead.

We then shortened and reset the rigging as before, after which we 'fished' the main-boom, repaired the mainsail, which had been badly torn, and also the jib. To repair the rudder, we fastened two ropes to the after part of the rudder blade, carrying one up either side of the vessel and fastening them to the tiller. The mizzen-mast, being in four pieces and therefore beyond repair, our work was completed at half-past four, and the *Sea Queen* was ready to sail again. Vincent, who had been attending to the cooking during the day, giving us a hand when called upon, sung out, 'Come down here, you sailormen, and have a number one and a half cup of coffee,' and as the weather was calm we joined the cook in the cabin and over a nice cup of coffee and a smoke decided to sail back to Yokohama, a distance of about three hundred and twenty miles.

Half an hour later a light breeze sprang up from the south-west accompanied by light drizzling rain, so we set our reefed mainsail and jib. Under these two sails, and with a moderate breeze, the little *Sea Queen* sailed along again almost as if nothing had happened. The rudder also worked as well as ever. The only trouble then remaining was the ever-present perfume of fish in our bedding and most of our clothes, which were saturated with the horrible oil, and since there was no prospect of eliminating the smell, we cast them all overboard.

15
A Love of Sensation

Erling Tambs, a Norwegian, left Norway with his wife, whom he had just married, in August 1928, on a voyage that was to be their honeymoon. On the very first night there was a 'heavy blow'. The *Teddy* had her fair share of bad weather in the Channel and Atlantic; but Tambs sailed on, through the Panama Canal and into the Pacific. Surely Tambs was foolhardy in the extreme, as in his own words, 'there were a few flaws in our equipment, we had practically no charts, we had neither instruments, nautical books, nor other navigational facilities, only an old card compass which ran wild whenever there was a bit of a sea going. We had no spare sails . . . But, among the various possessions at our disposal, there figured a bag of potatoes and a fishing line.'

The authorities in Norway tried to stop the voyage on this account, but found they had no authority over such a venture; so the *Teddy* left, never to return, for she was wrecked in the Pacific. Fortunately, Tambs, his wife, two children – yes, two were born on the voyage – and their dog were all saved.

The catalogue that Tambs gives of the equipment he lacked shows plainly enough that he knew what he should have had. Indeed he was trained in square-rigged ships and had sailed small boats single-handed. He was no novice.

I have included an extract from his account of his voyage because he writes of an electrical storm as no other seaman has done. I always remember too, this account of his being stranded on shore when his dinghy drifted away. His wife, not knowing why he returned to the yacht on a raft, through water abounding with sharks, thought it a queer outcrop of his love of sensation.

A. R.

from 'THE CRUISE OF THE TEDDY'
by ERLING TAMBS

THE Bay of Panama is a tedious enough place for a sailing vessel to navigate at all times of the year, but more particularly is it so in the rainy season. The prevailing wind – if such a term as prevailing can be applied at all – is SW. An entrance to the bay is generally easier to negotiate than an exit. However, the Bay of Panama and the waters west of it to Cocos Island and beyond are in reality an area of doldrums. A full description of the weather conditions of these regions would demand much time and patience from both writer and reader, but, speaking in general, the rule is light airs and calms alternating with thunderstorms and squalls from varying directions.

Around Cocos the unreliability of the conditions of navigation is further accentuated by unruly ocean currents, which change force and direction in a most unexpected manner.

It is, therefore, not in the least surprising that even the great seafarer Captain Dampier, having set out to determine the exact position of Cocos, failed to find the island. After cruising about for six weeks he finally retraced his course in the firm conviction that Cocos Island was a myth.

Certainly, when Captain Dampier set out, he was ignorant of the exact position of the island, but as to that it was only approximately that I knew my own position. And yet the idea of missing Cocos never entered my head. I was accustomed to rely on my luck. It was only after I had witnessed the mysterious way in which this island sometimes plays hide-and-seek, wrapping itself up in a cloud or hiding behind squalls, so that one might easily sail by at close range without any idea of its presence, that I realized my luck in finding it so readily.

The Bay of Panama and the adjoining waters are alive with fish of many varieties. Every little while *Teddy* would be surrounded by dense schools of them. A few I speared, others I caught with a short line, but no opportunity came my way of trying my luck on big-game fish. Before we rounded Cape Mala, our slow progress was not in favour of fishing with spinners, while afterwards matters of much greater moment occupied my attention.

This was rather unfortunate, however, seeing that sailfish abounded in these waters and huge skates, some of them surely weighing several tons, were jumping about and disporting themselves continually. Sometimes they made me jump too, as they fell back with a tremendous resounding splash into the sea.

Since there was hardly any wind except in the squalls, I had to make the most of them while they lasted. This meant running a certain risk, because one never knew what they might bring. The range of their possibilities lay between a gentle shower and a howling gale. Once or twice I lowered the main-sail before an approaching squall and rejoiced afterwards at having done so, but more frequently I was caught and had to make the best of conditions. This latter state of affairs generally occurred at night, when it was not so easy to define the nature of the squalls or even to see whence they were approaching. At times it was rather exciting.

But all these troubles pale into insignificance when I think of an electric storm we experienced during the night between April 28th and 29th. This storm was of such demonic violence and accompanied by so many uncanny side phenomena, that the bare thought of it is even yet exceedingly unpleasant to me.

The night, moonless and densely clouded, had settled around us with pitchy darkness. One could not see a hand held before the eyes. Rain came down in torrents. Repeatedly it drowned the riding light, until I abandoned the attempt of relighting it. Only the binnacle lamp was burning, throwing its faint rays on the cockpit combing, on a hand that was groping for a sheet rope or on a shining black oilskin coat.

And then the tempest broke loose. Crackling lightning bore down from out of greenish-brown poisonous-looking clouds that crowded low above the phosphorescent masthead. In ever quicker succession it swished down into the sea to right and left, some of the flashes in the immediate vicinity of the quivering boat, while the incessant roar of thunder, deafening the ear, shattering the nerves, sounded like hell let loose, like the infernal gunfire of a million gigantic demons at war.

The wind kept veering round from one direction to the other, blowing, sometimes, with hurricane force. I stood at the tiller tensely watching and running before the gale in order to save the rigging.

Then there would be a lull, intervals between the bursts of lightning – intervals of utter darkness.

I could see absolutely nothing. My only object was to keep the wind well aft and watch out that she did not jibe. Whatever I did, I must not jibe!

Gradually I could discern the greenish glow of those phosphorescent spar ends, until new flashes of lightning would make every detail distinctly visible in a ghastly bright light. I can still see that weird intensified picture of the deck asplash with the downpour, which came in hissing gushes, of the shiny black rigging and of the straining grey canvas, streaming with driving rain.

Then the horrible outburst would recontinue, fiercer, apparently, and more fiendish than ever. At times the whole sky would be a dense cobweb of lightning, flooding every crack and corner with an abominable brightness, and then again, in the blackness that followed, we would be rushing through wide streaks of dully glowing water, which stood out sharply, without surroundings, without background, like a flood of luminous milk in an empty space. What it was, I cannot tell. Perhaps the phosphorescence caused by billions of microscopic beings, isolated by meeting currents – perhaps only one more amongst many weird electric phenomena, which added to the indescribable horror of this night. More than once I almost expected the whole universe to explode.

According to my calculations, we should be close to Cocos Island. What if we ran on to a reef? That would bring the end, extinction. But what could I do? I had no will and no power but to obey the elements, rush onward, somewhere, anywhere. . . .

To oppose them would be fatal.

How utterly futile, however, it is to attempt to describe what one cannot even understand. What are words? What are epithets? Empty, surely, and without force or meaning

compared with the horror of that night, a night which gave me
a more vivid and terrifying picture of inferno than my imagination could ever have created. It was like a weird prelude to the
Day of Judgement. And when, at last, morning broke, when
the weather cleared up and the sun showed his face between
retreating clouds, my relief was unbounded.

Never has that mighty life-giver been greeted with greater
satisfaction.

After two more days of light and unsettled weather, we
sighted Cocos, a haystack resting on the northern horizon some
thirty miles off. At that moment it stood out very clearly between the neighbouring clouds. The day was fine, and a gentle
sou'westerly breeze carried us along towards our destination. By
noon we were abreast of the first off-lying rocks, sharp fang-like
pinnacles that protrude gleamingly from out of the depth, the
whitewashed resting-places of innumerable sea-birds.

As I had been warned that the chart of Cocos Island was not
too reliable, I determined, after just missing a patch of sunken
rocks off the south-eastern point, to keep well away from the
coast, until nearly abreast of Cone Rock. Then, turning to port,
we worked our way against the light breeze into the smooth
loneliness of beautiful Chatham Bay.

The water in the bay was clear as crystal, so extraordinarily
clear, indeed, that one might almost imagine the surface to
be only a plate-glass roof over the deeps below, where every
detail could be distinguished even down to fifteen fathoms or
more.

The bottom of the bay consisted of white sand with large
boulders of blue-black coral scattered about. *Teddy* clearly out-
lined, seemed to float unsupported above it, as if in air. Everywhere the sea was alive with fish of a hundred varieties, most of
them gaily coloured as if in carnival attire. Sharks too were
numerous wherever the eye scanned the water.

It had been my intention to anchor in some five fathoms, but
the transparency of that sea was so deceptive that when I

dropped the hook, I found that we were still in fully ten fathoms of water. If I had taken soundings, we might have gone much farther in and thereby saved much labour of rowing, but, having once dropped anchor, we remained where we were, some four or five hundred yards from the head of the bay.

The afternoon being well advanced, we decided to stay on board and spent the remaining hours of daylight in fishing and in watching the strange doings of those queer and fascinating dwellers of the deep. It was like looking into a huge and wonderful aquarium.

The next morning, collecting as many buckets and empty kerosene tins as our little dinghy would hold, I prepared to go ashore and fetch water from the brook at the head of the bay in order to replenish our water tanks. Thinking that I might gather a few coco-nuts at the same time, I also took a hatchet along.

Cheerfully setting off for my first landing on Treasure Island, I had scarcely come twenty yards away from *Teddy*, when I discovered that I had company, a huge shark following in my wake.

Sharks I regard as repulsive beasts, and I know them to be cunning. They never attack unless it is possible to do so in comparative safety, unless they have a decided advantage over their victim. I did not like this shark following me. I did not like his company.

I pulled harder; he kept pace easily, only, the faster I rowed, the more determined he became until, after a few more strokes, he disappeared under the dinghy.

Knowing how easily the little boat would capsize, I felt fairly uncomfortable but nevertheless continued rowing, until I felt the dinghy lifting slightly on one side, as the shark came up alongside. In desperate excitement I grabbed the hatchet and dealt the brute a slash in the back, which caused him to retreat in such a hurry that he nearly upset the boat after all.

I saw him no more, but a streak of blood in the clearness of the water showed me the course he had taken.

L

After this incident I never went rowing without the hatchet, and I took good care that it lay handy, especially when I was taking Julie and Tony with me in the boat.

It was only the next day, however, when the dinghy became the cause of another exciting experience.

I had gone ashore at daybreak to look for something to eat, and as there were no nuts on the Cocos palms along the beach, I decided to try to find some up the hillside. It was nearly high water when I landed, and the beach, which at low water is dry to the extent of some sixty yards, was entirely submerged.

Meantime I pulled the dinghy well up on to the pebbly bank beneath the verdure, and fastening a piece of good rope to the end of the painter, I tied her up to a tree that grew handy. Whereupon I went on my errand.

The hillside rises abruptly all round the bay. It is overgrown with a variety of trees, ferns and shrub which are interwoven with a wilderness of tough-limbed creepers and fallen trees in every stage of decay. I had to cut my way with the hatchet.

In the beginning I made but slow progress. The ascent was steep, the soil was slippery and the shrubbery proved hard to combat, but as I came higher up, conditions became easier.

Gaining the first hilltop, I found the forest reasonably open, less encumbered with creepers and underbrush. I began to enjoy my exploring.

The vegetation was a picturesque combination of Central American and Pacific island flora. There were tree ferns and palms of many kinds, banyan, pandanus and a hundred other trees, orchids and an abundance of large-leafed weeds.

The morning was still fresh and the air was fragrant with strange and pleasant odours. The birds were almost tame. Large butterflies fluttered about. Absorbed in the surroundings, I had long since forgotten what I came for. On and on I went, finding ever some new vista opening up for my admiration, another freak of nature to marvel at.

From time to time I would mark a tree with my hatchet to enable me to find my way back. Thus I blazed my trail.

In this manner perhaps two hours had elapsed before my conscience overcame my zeal to explore and caused me reluctantly to retrace my steps.

Returning was easy enough. I followed my own track, and after half an hour's brisk walking came to the hillslope, where an uncertain peep through the rustling foliage permitted me a faint glimpse of the bay and of *Teddy* at anchor.

But I imagined I saw something more. I did not stop to make sure, because that something would not let me. On the contrary, it made me rush down the hillside and through the thicket at a breakneck pace.

Presently I stood under the palms on the beach:

The *dinghy was gone*!

So it was indeed the dinghy that a little while ago I saw swinging round the point.

One glance showed me how it had happened; it was just after New Moon, the tide had risen to an unusual height; the surf reaching the light craft had set her afloat and then waves and tide had combined to saw off the painter against the sharp edge of a large rock. The rope with part of the painter was still as I had tied it.

Imagine my predicament!

Out there, some five hundred yards away lay *Teddy* at her anchor, with Tony and Julie on board.

She could do nothing. I had to get on board by my own means, and as soon as possible, while there was yet a chance to overtake the dinghy. Swim there? That would be madness. The bay was simply crowded with sharks; they were everywhere in scores, in hundreds, in thousands. Even in the shallow water along the beach, in less than two feet, their ugly back fins were abundantly evident above the surf.

I might succeed in climbing around to a steep cliff on the eastern side of the bay, but even then I would still have two or three hundred yards to swim, and that area abounded with big sharks.

These and a hundred other reflections flashed through my

mind in a fraction of the time it takes to relate them. Indeed, my plans were laid probably in less than a minute and immediately I set about putting them into operation. Fortunately I had at least a hatchet.

The receding tide had already uncovered a portion of the smooth sand. Dry driftwood lay in heaps just behind the beach. In a few minutes I had carried together a respectable pile. I did not have time to worry about the scorpions, which dropped liberally on to the sand from out of the burdens I carried.

Saplings were plentiful and close at hand. I cut down a score and tied them together into a frame by means of rope-yarns; the rope and what was left of the painter admirably served my purpose. Three saplings on either side and four at the bottom made a crate which I subsequently filled with light driftwood, endeavouring in spite of my haste, to place them so as to strengthen the frame. By the time my primitive raft was ready, I was soaked with perspiration. It ran into my eyes and blinded me.

And yet the worst of my labours were yet to come: I must launch this porcupine-like piece of shipbuilding. It resisted my efforts as if rammed into the ground. I jerked and pulled. I wept and cursed. Several times I was forced to lighten the thing and shove it into deeper water before refilling it and relashing the saplings which served as top beams. And then a sea would come along and shove the whole outfit on to the sand again.

However, success at last rewarded me; my porcupine was afloat and armed with my pole I managed to scramble aboard.

Having a fair idea of the set of the current, I judged that by poling myself along to the eastward, I should eventually get into a position whence the tide would carry me right down on to the *Teddy*.

At the outset things went fairly satisfactorily, although that awful craft made every attempt to fall to bits, but suddenly, without warning, the sea-bottom dropped away. I could no longer reach it with my pole, and I had not come nearly as far east as I should have come.

I tried to row, but that was evidently more than my raft would stand. Besides, rowing did not benefit me in the slightest.

The current carried me slowly on. Sharks were gathering around in large numbers. I could see them even between the branches on which I stood. Twigs kept continually floating away from the raft. They surrounded me in an ever-widening circle.

Teddy drew nearer: it was evident that I would drift by!

Julie was below, possibly asleep.

I began to hail the boat, shouting at the top of my voice. No one appeared.

This was a nice situation! I had certainly come from the frying-pan into the fire. If I drifted to sea in this frigate, my swan song would be sung.

Again I called, howling so loudly that it seemed as though the boughs beneath me rose up in protest and threatened to leave the assembly. Praise and Glory! I heard the dog barking below – how wonderfully sweet that gruff voice sounded! – and then Julie appeared.

She clasped her hands in consternation when she saw me. I was some forty yards from the boat, and I was about to drift by.

However, by great good luck, a heavy fishing line lay uncoiled on deck. Tied to one end it had a square piece of wood, on which I used to wind up the line before putting it away. With this the connexion was eventually established. And none too soon.

The rest was easy. Pulling myself very gently up to the *Teddy*, I caught hold of the main boom end and swung myself on board, just as my frail raft on the slight collision with the boat dissolved itself into a mass of floating driftwood.

My wife had at the time no idea of the reason for this 'queer outcrop of my love of sensation' as she termed it, but I gave her the essentials of the case in a few words. Ten minutes later I had put a buoy on to the chain, hoisted the sails, and we went to sea in pursuit of a runaway dinghy from Ancas Boatbuilding Yard at Arendal.

Following the direction of wind and current as best we could for two hours, we were just on the point of giving up the chase and returning, when at last we caught sight of it. By this time it was half full of water, but fortunately none of the gear in her had as yet been lost.

Before night we were again at anchor in Chatham Bay. The dinghy floated astern, tied up with a new painter. The incident had after all cost us nothing but a bit of rope, a scare and a good deal of work, which latter, I am told, is but wholesome.

However, Spare Provisions [their dog] was that night over-fed in spite of our shortage of supplies.

16
Man Overboard

In 1929 the full-rigged Finnish ship *Grace Harwar* took part in the grain race from Australia to Europe. She was plagued by bad weather for the whole of the passage to Cape Horn, which took two months instead of the usual four or five weeks. This extract is taken from an account of the voyage, published in *The Times* of September 5th, 1929.

A. R.

from 'THE TIMES'

SHORTLY before getting to Cape Horn the *Grace Harwar* was heavily beset by a series of gales which did much damage on her decks; at one stage it was necessary to put oil on the seas to break their force. In a gale in the depth of winter in the South Atlantic the ship sprang a leak, which fortunately proved not to be serious, although it was necessary to man the pumps more or less regularly for the remainder of the voyage.

After this gale, while big seas still swept the decks, one of the Finnish members of the crew went overboard from the main braces. He was seen and a lifebuoy thrown, which he was able to secure. At the time the lifeboats were lashed down for security in order that they might not be washed away; no falls were rove in the davits, and it was dangerous to bring the ship to the wind before which she was running heavily under a big press of sail. She was brought into the wind without hesitation, however; the lifeboat lashings were cut with an axe, falls rove off rapidly, the lee boat was swung out, and volunteers were called for.

It was late in the afternoon then, in the depth of winter, and there was a threat of renewed gale in the sky. Without hesitation the whole of the crew responded. Six were chosen – four Swedish Finns, a Frenchman, and an Australian – and the boat put off.

No one then knew where the missing man had gone. He could not be seen from the ship, and in the boat the seas were so big that, falling into their troughs, it was impossible to see beyond one of them, and even on the crests for the greater time the horizon was bounded by the third sea. Darkness was coming down; no one knew in which direction to steer. The mate tried to steer a course against the wind, remembering that the ship had been running before it, but its direction changed. He tried, in desperation, to follow some circling albatrosses, in the hope that they were circling over the floating figure. They came and circled over the boat. He tried to follow back what was left of the ship's wake, but that was hopeless. A sailing ship, even running heavily, leaves little wake. Darkness was coming on quickly, and the threat of storm had grown to certainty; in the last moment of light, when everybody had given up hope, the lifebuoy was seen and the man was found. He was unconscious then and suffered badly from exposure, but after some weeks was able to return to duty as if nothing had happened.

It was a curious thing that the boat had been headed all the time in the direction of the lifebuoy, but none of her crew had known it.

17
Pitchpoled

From time to time seamen meet fresh waves which are of such size that nothing can prevent a small craft from being laid right over on her nose or even turned completely over through 360°. These exceptional waves have been encountered by ships in all parts of the world, including the Mediterranean, although most often they are reported from the Southern Ocean. They probably result from a number of underlying but different wave motions which coincide to build up the super wave. Meeting such a wave cannot be regarded as a rare and unlikely event, but as a real possibility.

Miles Smeeton and his wife, with John Guzzwell, met such a wave in the South Pacific when they were heading for the Horn in *Tzu Hang*. Remarkably they saved their dismasted and waterlogged yacht by making temporary repairs and a jury rig to sail her into a Chilean harbour.

A. R.

from '*ONCE IS ENOUGH*' by MILES SMEETON

IT must have been nearly five in the morning, because it was light again, when the noise of the headsails became so insistent that I decided to take in sail. I pulled on my boots and trousers. Now that I had decided to take some action I felt that it was already late, and was in a fever to get on with it. When I was dressed I slid back the hatch, and the wind raised its voice in a screech as I did so. In the last hour there had been an increase in the wind, and the spindrift was lifting, and driving across the face of the sea.

I shut the hatch and went forward to call Beryl. She was awake, and when I went aft to call John, he was awake also. They both came into the doghouse to put on their oilies. As we got dressed there was a feeling that this was something unusual;

it was rather like a patrol getting ready to leave, with the
enemy in close contact. In a few minutes we were going to be
struggling with this gale and this furious-looking sea, but for the
time being we were safe and in shelter.

'Got your life-lines?' Beryl asked.

'No, where the hell's my life-line? It was hanging up with the
oilies.' Like my reading glasses, it was always missing.

'Here it is,' John said. He was buckling on a thick leather belt
over his jacket, to which his knife, shackle spanner, and life-
line were attached. His life-line was a thin nylon cord with a
snap-hook at the end, and Beryl's, incongruously, was a thick
terylene rope, with a breaking strain of well over a ton.

'Got the shackle spanner?' I asked. 'Never mind, here's a
wrench. Is the forehatch open?' Someone said that they'd
opened it.

'Beryl, take the tiller. John and I'll douse the sails. Come on
boys, into battle.' I slid the hatch back again and we climbed up
one after the other. We were just on the crest of a wave and
could look around over a wide area of stormy greyish-white sea.
Because we were on the top of a wave for a moment, the seas did
not look too bad, but the wind rose in a high pitched howl, and
plucked at the double shoulders of our oilies, making the flaps
blow up and down.

The wave passed under *Tzu Hang*. Her bowsprit rose, and she
gave a waddle and lift as if to say, 'Be off with you!' Then the
sea broke, and we could hear it grumbling away ahead of us,
leaving a great wide band of foam behind it.

Beryl slipped into the cockpit and snapped her life-line on to
the shrouds. John and I went forward, and as we let go of the
handrail on the doghouse, we snapped the hooks of our life-lines
on to the rail, and let them run along the wire until we had hold
of the shrouds. The wind gave us a push from behind as we
moved. I went to the starboard halliard and John to the port,
and I looked aft to see if Beryl was ready. Then we unfastened
the poles from the mast and let the halliards go, so that the sails
came down together, and in a very short time we had them

secured. We unhooked them from the stays, bundled them both down the forehatch, and secured the two booms to the rails. As we went back to the cockpit, we were bent against the pitch of the ship and the wind. Beryl unfastened the sheets from the tiller and we coiled them up and threw them below.

'How's she steering?'

'She seems to steer all right, I can steer all right.'

'We'll let the stern line go anyway, it may be some help.'

John and I uncoiled the 3-inch hawser, which was lashed in the stern, and paid it out aft. Then we took in the log-line, in case it should be fouled by the hawser. By the time everything was finished, my watch was nearly due, so I took over the tiller from Beryl, and the others went below. The hatch slammed shut, and I was left to myself. I turned my attention to the sea.

The sea was a wonderful sight. It was as different from an ordinary rough sea, as a winter's landscape is from a summer one, and the thing that impressed me most was that its general aspect was white. This was due to two reasons: firstly because the wide breaking crests left swathes of white all over the sea, and secondly because all over the surface of the great waves themselves, the wind was whipping up lesser waves, and blowing their tops away, so that the whole sea was lined and streaked with this blown spume, and it looked as if all the surface was moving. Here and there, as a wave broke, I could see the flung spray caught and whirled upwards by the wind, which raced up the back of the wave, just like a whirl of wind-driven sand in the desert. I had seen it before, but this moving surface, driving low across a sea all lined and furrowed with white, this was something new to me, and something frightening, and I felt exhilarated with the atmosphere of strife. I have felt this feeling before on a mountain, or in battle, and I should have been warned. It is apt to mean trouble.

For the first time since we entered the Tasman there were no albatrosses to be seen. I wondered where they had gone to, and supposed that however hard the wind blew it could make no difference to them. Perhaps they side-slipped out of a storm area, or perhaps they held their position as best they could until the storm passed, gliding into the wind and yet riding with the storm until it left them.

I kept looking aft to make sure that *Tzu Hang* was dead stern on to the waves. First her stern lifted, and it looked as if we were sliding down a long slope into the deep valley between this wave and the one that had passed, perhaps twenty seconds before; then for a moment we were perched on the top of a sea, the wind force rose, and I could see the white desolation around me. Then her bowsprit drove into the sky, and with a lurch and a shrug, she sent another sea on its way. It was difficult to estimate her speed, because we had brought the log in, and the state of the water was very disturbed, but these waves were travelling a great deal faster than she was, and her speed seemed to be just sufficient to give her adequate steerage way, so that I could correct her in time to meet the following wave.

Suddenly there was a roar behind me and a mass of white water foamed over the stern. I was knocked forward out of the cockpit on to the bridge deck, and for a moment I seemed to be sitting in the sea with the mizzen-mast sticking out of it upright beside me. I was surprised by the weight of the water, which had burst the canvas windscreen behind me wide open, but I was safely secured by my body-line to the after shroud. I scrambled back into the cockpit and grabbed the tiller again, and pushed it hard over, for *Tzu Hang* had swung so that her quarter was to the sea. She answered slowly but in time, and as the next sea came up, we were stern on to it again. The canvas of the broken windbreak lashed and fluttered in the wind until its torn ends were blown away.

Now the cloud began to break up and the sun to show. I couldn't look at the glass, but I thought that I felt the beginning of a change. It was only the change of some sunlight, but the sunlight seemed to show that we were reaching the bottom of this depression. Perhaps we would never get a chance again to film such a sea, in these fleeting patches of brilliance. I beat on the deck above John's bunk and called him up. I think that he had just got to sleep, now that the sails were off her, and there was someone at the helm. I know that I couldn't sleep before. He looked sleepy and disgruntled when he put his head out of the hatch.

'What about some filming, John?'

'No, man, the sea never comes out.'

'We may never get a sea like this again.'

'I don't want to get the camera wet, and there's not enough light.'

'No, look, there's a bit of sun about.'

As he was grumbling, like an old bear roused out of its winter quarters, he looked aft and I saw his expression change to one of interest.

'Look at this one coming up,' he said, peering over the top of the washboards, just the top of his head and his eyes showing. 'Up she goes,' he ducked down as if he expected some spray to come over, and then popped his head up again. 'Wait a minute,' he said, 'I'll fix something up,' and he slammed the hatch shut and disappeared below again.

He came up in a few minutes, fully equipped. He had the camera in a plastic bag with the lens protruding through a small hole. He took some shots. The lens had to be dried repeatedly, but the camera was safe in its bag, and we had no more wave tops on board. Presently he went down again.

John relieved me for breakfast, and when I came up it seemed to be blowing harder than ever.

'How's she steering?' I asked him.

'Not bad,' he said. 'I think she's a bit sluggish, but she ought to do.'

I took over again, and he went below; no one wanted to hang about in this wind. I watched the sixty fathoms of 3 inch hawser streaming behind. It didn't seem to be making a damn of difference, although I suppose that it was helping to keep her stern on to the seas. Sometimes I could see the end being carried forward in a big bight on the top of a wave. We had another sixty fathoms, and I considered fastening it to the other and streaming the two in a loop, but I had done this before, and the loop made no difference, although the extra length did help to slow her down. We had oil on board, but I didn't consider the emergency warranted the use of oil. For four hours now we had been running before this gale, running in the right direction, and we had only had one breaking top on board, and although I had been washed away from the tiller, *Tzu Hang* had shown little tendency to broach to. To stop her and to lie a-hull in this big sea seemed more dangerous than to let her run, as we were doing now. It was a dangerous sea I knew, but I had no doubt that she would carry us safely through and as one great wave after another rushed past us, I grew more and more confident.

Beryl relieved me at nine o'clock. She looked so gay when she came on deck, for this is the sort of thing that she loves. She was wearing her yellow oilskin trousers and a yellow jumper with a hood, and over all a green oilskin coat. So that she could put on enough pairs of socks, she was wearing a spare pair of John's sea-boots. She was wearing woollen gloves, and she had put a plastic bag over her left hand, which she wouldn't be using for the tiller. She snapped the shackle of her body-line on to the shroud, and sat down beside me, and after a minute or two she took over. I went below to look at the glass and saw that it had moved up a fraction. My camera was in the locker in the doghouse, and I brought it out and took some snaps of the sea. Beryl was concentrating very hard on the steering. She was looking at the compass, and then aft to the following sea, to make sure that she was stern on to it, and then back to the compass again, but until she had the feel of the ship she would trust more to the compass for her course than to the wind and the waves. I took one or two snaps of Beryl, telling her not to look so serious, and to give me a smile. She laughed at me.

'How do you think she's steering?'

'Very well, I think.'

'We could put the other line out. Do you think she needs it? The glass is up a bit.'

'No, I think she's all right.'

'Sure you're all right?'

'Yes, fine, thanks.'

I didn't want to leave her and to shut the hatch on her, and cut her off from us below, but we couldn't leave the hatch open, and there was no point in two of us staying on deck. I took off my oilskins, put the camera back in its plastic bag in the locker, and climbed up into my bunk. The cat joined me and sat on my stomach. She swayed to the roll and purred. I pulled my book out of the shelf and began to read. After a time, I heard John open the hatch again and start talking to Beryl. A little later he went up to do some more filming. As the hatch opened there was a roar from outside, but *Tzu Hang* ran on straight and true, and I felt a surge of affection and pride for the way she was doing. 'She's a good little ship, a good little ship,' I said to her aloud, and patted her planking.

I heard the hatch slam shut again, and John came down. He went aft, still dressed in his oilskins, and sat on the locker by his bunk, changing the film of his camera. Beneath him, and lashed securely to ring-bolts on the locker, was his tool-box, a large wooden chest, about 30 inches by 18 inches by 8 inches, crammed full with heavy tools.

My book was called *Harry Black*, and Harry Black was following up a wounded tiger, but I never found out what happened to Harry Black and the tiger.

When John went below, Beryl continued to steer as before, continually checking her course by the compass, but steering more by the wind and the waves. She was getting used to them now, but the wind still blew as hard as ever. In places the sun broke through the cloud, and from time to time she was in sunshine. A wave passed under *Tzu Hang*, and she slewed slightly. Beryl corrected her easily, and when she was down in the hollow she looked aft to check her alignment. Close behind her a great wall of water was towering above her, so wide that she couldn't see its flanks, so high and so steep that she knew *Tzu*

Hang could not ride over it. It didn't seem to be breaking as the other waves had broken, but water was cascading down its front, like a waterfall. She thought, 'I can't do anything, I'm absolutely straight.' This was her last visual picture, so nearly truly her last, and it has remained with her. The next moment she seemed to be falling out of the cockpit, but she remembers nothing but this sensation. Then she found herself floating in the sea, unaware whether she had been under water or not.

She could see no sign of *Tzu Hang*, and she grabbed at her waist for her life-line, but felt only a broken end. She kicked to tread water, thinking, 'Oh, God, they've left me!' and her boots, those good roomy boots of John's, came off as she kicked. Then a wave lifted her, and she turned in the water, and there was *Tzu Hang*, faithful *Tzu Hang*, lying stopped and thirty yards away. She saw that the masts were gone and that *Tzu Hang* was strangely low in the water, but she was still afloat and Beryl started to swim towards the wreckage of the mizzen-mast.

As I read, there was a sudden, sickening sense of disaster. I felt a great lurch and heel, and a thunder of sound filled my ears. I was conscious, in a terrified moment, of being driven into the front and side of my bunk with tremendous force. At the same time there was a tearing cracking sound, as if *Tzu Hang* was being ripped apart, and water burst solidly, ranging into the cabin. There was darkness, black darkness, and pressure, and a feeling of being buried in a debris of boards, and I fought wildly to get out, thinking *Tzu Hang* had already gone. Then suddenly I was standing again, waist deep in water, and floorboards and cushions, mattresses and books, were sloshing in wild confusion round me.

I knew that some tremendous force had taken us and thrown us like a toy, and had engulfed us in its black maw. I knew that no one on deck could have survived the fury of its strength, and I knew that Beryl was fastened to the shrouds by her life-line, and could not have been thrown clear. I struggled aft, fearing what I expected to see, fearing that I would not see her alive again. As I went I heard an agonized yell from the cat, and thought, 'Poor thing, I cannot help you now.' When I am angry, or stupid and spoilt, or struggling and in danger, or in distress, there is a part of me which seems to disengage from my body, and to survey the scene with a cynical distaste. Now that I was afraid, this other half seemed to see myself struggling through all the floating debris, and to hear a distraught voice crying, 'Oh God, where's Bea, where's Bea?'

As I entered the galley, John's head and shoulders broke water by the galley stove. They may have broken water already, but that was my impression anyway. John himself doesn't know how he got there, but he remembers being thrown forward from where he was sitting and to port, against the engine exhaust and the petrol tank. He remembers struggling against the tremendous force of water in the darkness, and wondering how far *Tzu Hang* had gone down and whether she could ever get up again. As I passed him he got to his feet. He looked sullen and obstinate, as he might look if someone had offended him, but he said nothing. There was no doghouse left. The corner posts had been torn from the bolts in the carlins, and the whole doghouse sheared off flush with the deck. Only a great gaping square hole in the deck remained.

As I reached the deck, I saw Beryl. She was thirty yards away on the port quarter on the back of a wave, and for the moment above us, and she was swimming with her head well out of the water. She looked unafraid, and I believe that she was smiling.

'I'm all right, I'm all right,' she shouted.

I understood her although I could not hear the words, which were taken by the wind.

The mizzen-mast was in several pieces, and was floating between her and the ship, still attached to its rigging, and I saw that she would soon have hold of it. When she got there, she pulled herself in on the shrouds, and I got hold of her hand. I saw that her head was bleeding, and I was able to see that the cut was not too serious but when I tried to pull her on board, although we had little freeboard left, I couldn't do it because of

the weight of her sodden clothes and because she seemed to be
unable to help with her other arm. I saw John standing amidships. Incredibly he was standing, because, as I could see now,
both masts had gone, and the motion was now so quick that I
could not keep my feet on the deck. He was standing with his
legs wide apart, his knees bent and his hands on his thighs. I
called to him to give me a hand. He came up and knelt down
beside me, and said, 'This is it, you know, Miles.'

But before he could get hold of Beryl, he saw another wave
coming up, and said, 'Look out, this really is it!'

Beryl called, 'Let go, let go!'

But I wasn't going to let go of that hand, now that I had got
it, and miraculously *Tzu Hang*, although she seemed to tremble
with the effort, rode another big wave. She was dispirited and
listless, but she still floated. Next moment John caught Beryl by
the arm, and we hauled her on board. She lay on the deck for a
moment, and then said, 'Get off my arm, John, I can't get up.'

'But I'm not on your arm,' he replied.

'You're kneeling on my arm, John.'

'Here,' he said, and gave her a lift up. Then we all turned on
our hands and knees, and held on to the edge of the big hole in
the deck.

Up to now my one idea had been to get Beryl back on board,
with what intent I do not really know, because there was so
much water below that I was sure *Tzu Hang* could not float
much longer. I had no idea that we could save her, nor, John
told me afterwards, had he. In fact, he said, the reason why he
had not come at once to get Beryl on board again, was that he
thought *Tzu Hang* would go before we did so. After this first
action, I went through a blank patch, thinking that it was only a
few moments, a few minutes of waiting, thinking despondently
that I had let Clio down. Beryl's bright, unquenchable spirit
thought of no such thing. 'I know where the buckets are,' she
said. 'I'll get them!'

This set us working to save *Tzu Hang*.

Beryl slipped below, followed by John, but for the time being

I stayed on deck and turned to look at the ruins that had been *Tzu Hang*. The tiller, the cockpit coaming, and every scrap of the doghouse had gone, leaving a 6 foot by 6 foot gap in the deck. Both masts had been taken off level with the deck, the dinghies had gone, and the cabin skylights were sheared off a few inches above the deck. The bowsprit had been broken in two. The rail stanchions were bent all over the place, and the wire was broken. A tangle of wire shrouds lay across the deck, and in the water to leeward floated the broken masts and booms – the masts broken in several places. The compass had gone, and so had the anchor which had been lashed to the foredeck.

There could be no more desolate picture. The low-lying, water-logged, helpless hull, the broken spars and wreckage, that greyish-white sea; no bird, no ship, nothing to help, except that which we had within ourselves. Now the sun was gone again, the spindrift still blew chill across the deck, and the water lipped on to it, and poured into the open hull.

I think both John and I had been numbed with shock, but he recovered first and was working in a fury now, and a hanging cupboard door, some floorboards, and the Genoa erupted on to the deck. I hung on to them, so that they would not be blown or washed off, while he went down again for his tools. He found his tool-box jammed in the sink, and when he groped under water for the tin of galvanized nails in the paint locker, he found them on his second dip. I had intended to help John, as the first essential seemed to be to get the hole covered up, but he was working so fast, so sure now of what he was going to do, his mouth set in a grim determination, oblivious of anything but the work in hand, that I saw he would do as well without me, and now Beryl, who was trying to bale, found that she could not raise the bucket to empty it. I climbed down to where Beryl was standing on the engine, the water washing about her knees. We had to feel for some foothold on the floor-bearers because the engine cover had gone. A 70-lb keg of waterlogged flour floated up to us.

'Overboard with it,' I said.

'Mind your back,' said Beryl, as if we were working on the farm.

I picked it up, and heaved it on deck, Beryl helping with her good arm, but it seemed light enough, and John toppled it over into the sea, out of his way. Anything to lighten the ship, for she was desperately heavy and low in the water.

'We'd better bale through the skylight,' Beryl said, 'I'll fill the bucket, and you haul it up. We'll need a line. Here, take my life-line.' She undid her line from her waist and handed it to me, and I noticed that the snap-hook was broken. I tied it on to the handle of the plastic bucket.

We waded into the cabin feeling with our feet, because there were no floorboards to walk on. I climbed on to the bunk and put my head and shoulders through the skylight. Beryl was on the seat below with the water still round her knees. She filled the bucket and I pulled it up and emptied it, and dropped it down through the skylight again. It would have floated if she had not been there to fill it. It was the best that we could do, and although we worked fast, to begin with, we could just keep pace with the water coming in. No heavy seas broke over the ship and when a top splashed over, I tried to fill the aperture with my body as best I could.

John was doing splendidly. He had made a skeleton roof over the hatch with the door and floorboards, and nailed it down, and he had made it higher in the middle, so that it would spill the water when the sail was nailed over. He was nailing the folded sail over it now, using pieces of wood as battens to hold it down. It was a rush job, but it had to be strong enough to hold out until the sea went down. As soon as he had finished he went to the other skylight and nailed the storm-jib over it. Beryl and I baled and baled. As the bucket filled she called 'Right!' and I hauled it up again. Her voice rang out cheerfully from below, 'Right . . . Right . . . Right!' and John's hammer beat a steady accompaniment. *Tzu Hang* began to rise slowly, and at first imperceptibly, in the water.

When John had finished with the skylight, he called to me to ask if he should let the rigging screws go, so that the broken spars would act as a sea-anchor. I told him to do so, and he then went round the deck and loosened all the rigging-screws, leaving only one of the twin forestays attached to the deck. He had not much to work with, as the topmast forestay and the jibstay had gone with the mast, the forestay had smashed the deck fitting, and the other twin forestay had pulled its ringbolt through the deck, stripping the thread on the ringbolt. All the rigging was connected in some way or other, so now *Tzu Hang* drifted clear of her spars and then swung round, riding head to wind, on her single forestay. This forestay was attached to a mast fitting, on the broken mast, and the fitting was not equal to the strain now put on it, and it carried away. *Tzu Hang* swung away and drifted downwind, sideways to the sea, and that was the last that we saw of our tall masts, and the rigging, and the sails on the booms.

We baled and baled. We had two pumps on board, but the water that we were baling was filled with paper pulp from books and charts and labels. They would have clogged up with two strokes, and to begin with the pump handles themselves had been under water. John was now in the forecabin, standing on the bunks and baling through the skylight. Both he and Beryl were wearing the oilskins that they had been wearing when we upset, but I was still only in a jersey, and was beginning to feel very cold. I was continually wet with spray and salt water and lashed by the bitter wind, and my eyes were so encrusted and raw with salt, that I was finding it difficult to see. A broken spinnaker-pole rolled off the deck, showing that *Tzu Hang* was coming out of the water again, and was getting more lively. I saw a big bird alight by it and start pecking at it, and I supposed that this was also a sign that the wind was beginning to abate. I peered through rimy eyes to try and identify it, and saw that it was a giant fulmar. It was the first and only one that we saw.

After a time I became so cold that I could no longer pull up the bucket, in spite of Beryl's encouragement from below.

'This is survival training, you know,' she said.

It was the first joke. Survival training or no, I had to go below for a rest and we called a halt, and John came back from the main cabin and we sat on the bunk for a short time. Beryl found a tin of Horlicks tablets in a locker that had not been burst open, and she pulled off one of her oilskins and gave it to me.

'Where is Pwe?' I asked. 'Anyone seen her?'

'No, but I can hear her from time to time. She's alive. We can't do anything about her now.'

We were making progress, for the top of the engine was showing above the water. There was over a foot less in the ship already, and we were getting down to the narrower parts of her hull. We went back to work, and I found that now I had some protection from the wind the strength came back to my arms and I had no further difficulty. All through the day and on into the evening we baled, with occasional breaks for rest and more Horlicks tablets. Before dark, almost twelve hours after the smash, we were down below the floor-bearers again. After some difficulty we managed to get the primus stove, recovered from the bilge, to burn with a feeble impeded flame, and we heated some soup, but Beryl wouldn't have any. Now that the struggle was over for the time being, she was in great pain from her shoulder, and she found that she couldn't put her foot to the ground into the bargain. She had injured it stumbling about in the cabin, with no floorboards. Some blood-clotted hair was stuck to her forehead. Like a wounded animal, she wanted to creep into some dark place and to sleep until she felt better.

We found a bedraggled rag of a cat, shivering and cold, in the shelf in the bow, and with her, the three of us climbed into John's bunk to try and get some warmth from each other.

'You know,' I said, 'if it hadn't been for John, I think that we wouldn't have been here now.'

'No,' he said. 'If there hadn't been three of us, we wouldn't be here now.'

'I don't know,' said Beryl. 'I think you were the man, the way you got those holes covered.'

'I think my tool-box having jammed in the sink, and finding those nails is what saved us, at least so far.'

Beryl said, 'If we get out of this, everyone will say that we broached, but we didn't. They'll say that there was a woman at the helm.'

'If they know you they'll say that's how they know we didn't broach, and anyway, we didn't: we just went wham. Let's leave it.'

'How far are we off shore, Miles?' John asked.

'About 900 miles from the entrance to Magellan I should say.'

'If we get out of this, it will be some journey. If there is a lull tomorrow, I'll fix these covers properly.'

'Get her seaworthy again, and get her cleaned up inside. I think that's the first thing to do.'

Beryl was restless with pain and couldn't sleep. In the end we went back to our own soggy bunks. As we lay, sodden and shivering, and awoke from fitful slumber to hear the thud of a wave against the side of the ship and the patter and splash of the spray on the deck, we could hear the main-sheet traveller sliding up and down on its horse. The sheet had gone with the boom, but the traveller was still there, and this annoying but familiar noise seemed to accentuate the feeling that the wreck was just a dream in spite of the water which cascaded from time to time through the makeshift covers. The old familiar noises were still there, and it was hard to believe that *Tzu Hang* was not a live ship, still running bravely down for the Horn.

For several days to come, although all our energies were spent in overcoming the difficulties of the changed situation, it seemed impossible to accept its reality.

18
Uncharted Shoals

A previously unknown shoal or rock that has been sighted by a vessel, but the exact position of which is doubtful, is called a vigia. Many of these were reported in the old days of sail, and they were shown on the charts by dotted lines until their existence was confirmed or disproved by an accurate survey.

As sailing ships normally try to avoid shoal water these sightings were seldom supported by any soundings to give the actual depth.

When survey vessels have gone to check on the reports, frequently no trace has been found of the supposed danger. However, the crews of these ships themselves often sighted supposed shoals, but when they went to examine them, they found the sea to be discoloured by marine life or lava dust, with no shallow water. Waves breaking over rocks or shallows, another common report, were found to be shoals of fish jumping, or tide rips caused by the meeting of powerful currents.

But not all vigias are imaginary, as the *Maria Augusta* found when she was 500 miles from the nearest land. Noticing a change in the colour of the ocean she took soundings and found herself in less than ten fathoms.

Most seamen have enough anxiety from known dangers and uncertain weather, without going out of their way to visit vigias, whose position and extent are uncertain. None the less, this is what Conor O'Brien did, and here is a description of a vigia he encountered in the South Atlantic.

A. R.

from 'ACROSS THREE OCEANS'
by CONOR O'BRIEN

WHEN we sailed, the wind was, for the Falkland Islands, unexpectedly light and variable, with calms and fogs; during the first week there was only one day of a decent sailing breeze. I was now so near home that I was beginning to get anxious about the stories I should have to invent when I got there. The material was very scanty. I had seen no fabulous monsters, and I have rather a good eye for fabulous monsters; I saw the Lammergeyer in Switzerland in 1900 and the great sea serpent off the Kerry coast five years earlier. I had discovered no new lands; I sailed across a patch in the middle of the South Pacific marked 'Shoal' on the chart, but I saw no shoal; small wonder, for the dotted line that defined it enclosed an area as big as Sussex. I was now very near a more authentic shoal, one with an acknowledged position and a name though with no depth assigned to it; and I intended, if the weather and the time of day suited, to get a cast of the lead on it and so, perhaps, my name into print. But I found that I was likely to be thereabouts at three o'clock in the morning, which is not a good time to approach unsounded rocks, and I did not want to spoil a good day's run by stopping to look for it – after all it was very small game – and I passed it by. I was 60 miles past it at noon on the 6th of March, when the look-out reported two patches of kelp ahead. I passed them by. I considered that I was 600 miles from land, out of the track of vessels, and had no boat; and that the prudent man would run down to leeward of them, and if he wanted to investigate work back cautiously rather than run over them with all kites set. I kept passing by patches of kelp for about an hour, and then I saw that a swell, which I had not noticed before, was getting up; and right ahead of us it ran into a tall pyramid, and I need not say I lost no time in altering course; and the next swell that came along was like a steep hipped roof and had a little crest of white on top; and the next was light brown in colour with a pale spot in the middle, and why it did not break I do not know, but I take no chances with blind rollers and was out of danger by that time. Now if anyone wants to take soundings within a cable's length of Saoirse Rock he had better wait for finer weather with less sea and swell than I had, for I doubt if there is three fathoms of water over it; and he will have to wait a mighty long time. I made the best of my way out of the neighbourhood. I did indeed pass over some kelp, but I could see no bottom below it, and kelp may grow in

N

thirty fathoms of water. I wrote correctly on one page of my log-book that the latitude was 43° 21′ S. and the longitude uncertain, and on the opposite page incorrectly that the latitude was 40° 21′; and communicated the latter figure to the Hydrographer of the Navy. And thereby hangs a tale.

19
Taking to the Boats

At the back of his mind every sailor knows that one day he could suffer the loss of his ship. Few, who have had this experience and been forced to take to the boats in mid-ocean, have been able to conceal their distress as they saw their ship go down. Yet the calamity may suddenly give a young man his first command, a cockle-shell of a boat perhaps, yet a craft in which he is entirely responsible for the safety of those aboard. It is very difficult for small boats to keep together, especially in rough weather, and that is when the decisions of the man in charge count for much.

Here Conrad describes how youth turned a marine disaster into a stirring adventure.

A. R.

from '*YOUTH*' by JOSEPH CONRAD

THE skipper lingered disconsolately, and we left him to commune alone for a while with his first command. Then I went up again and brought him away at last. It was time. The ironwork on the poop was hot to the touch.

Then the painter of the long-boat was cut, and the three boats, tied together, drifted clear of the ship. It was just sixteen hours after the explosion when we abandoned her. Mahon had charge of the second boat, and I had the smallest – the 14-foot thing. The long-boat would have taken the lot of us; but the skipper said we must save as much property as we could – for the underwriters – and so I got my first command. I had two men with me, a bag of biscuits, a few tins of meat, and a breaker of water. I was ordered to keep close to the long-boat, that in case of bad weather we might be taken into her.

And do you know what I thought? I thought I would part company as soon as I could. I wanted to have my first command all to myself. I wasn't going to sail in a squadron if there were a chance for independent cruising. I would make land by myself. I would beat the other boats. Youth! All youth! The silly, charming, beautiful youth.

But we did not make a start at once. We must see the last of the ship. And so the boats drifted about that night, heaving and setting on the swell. The men dozed, waked, sighed, groaned. I looked at the burning ship.

Between the darkness of earth and heaven she was burning fiercely upon a disc of purple sea shot by the blood-red play of gleams; upon a disc of water glittering and sinister. A high, clear flame, an immense and lonely flame, ascended from the ocean, and from its summit the black smoke poured continuously at the sky. She burned furiously, mournful and imposing like a funeral pile kindled in the night, surrounded by the sea, watched over by the stars. A magnificent death had come like a grace, like a gift, like a reward to that old ship at the end of her laborious days. The surrender of her weary ghost to the keeping of stars and sea was stirring like the sight of a glorious triumph. The masts fell just before daybreak, and for a moment there was a burst and turmoil of sparks that seemed to fill with flying fire the night patient and watchful, the vast night lying silent upon the sea. At daylight she was only a charred shell, floating still under a cloud of smoke and bearing a glowing mass of coal within.

Then the oars were got out, and the boats forming in a line moved round her remains as if in procession – the long-boat leading. As we pulled across her stern a slim dart of fire shot out viciously at us, and suddenly she went down, head first, in a great hiss of steam. The unconsumed stern was the last to sink; but the paint had gone, had cracked, had peeled off, and there were no letters, there was no word, no stubborn device that was like her soul, to flash at the rising sun her creed and her name.

We made our way north. A breeze sprang up, and about noon all the boats came together for the last time. I had no mast or sail in mine, but I made a mast out of a spare oar and hoisted a

boat-awning for a sail, with a boat-hook for a yard. She was
certainly over-masted, but I had the satisfaction of knowing that
with the wind aft I could beat the other two. I had to wait for
them. Then we all had a look at the captain's chart, and, after a
sociable meal of hard bread and water, got our last instructions.
These were simple: steer north, and keep together as much as
possible. 'Be careful with that jury-rig, Marlow,' said the captain; and Mahon, as I sailed proudly past his boat, wrinkled his
curved nose and hailed, 'You will sail that ship of yours under
water, if you don't look out, young fellow.' He was a malicious
old man – and may the deep sea where he sleeps now rock him
gently, rock him tenderly to the end of time!

Before sunset a thick rain-squall passed over the two boats,
which were far astern, and that was the last I saw of them for a
time. Next day I sat steering my cockle-shell – my first command – with nothing but water and sky around me. I did sight
in the afternoon the upper sails of a ship far away, but said
nothing, and my men did not notice her. You see I was afraid
she might be homeward bound, and I had no mind to turn back
from the portals of the East. I was steering for Java – another
blessed name – like Bankok, you know. I steered many days.

I need not tell you what it is to be knocking about in an open
boat. I remember nights and days of calm when we pulled, we
pulled, and the boat seemed to stand still, as if bewitched within
the circle of the sea horizon. I remember the heat, the deluge of
rain-squalls that kept us baling for dear life (but filled our
water-cask), and I remember sixteen hours on end with a
mouth dry as a cinder and a steering-oar over the stern to keep
my first command head on to a breaking sea. I did not know
how good a man I was till then. I remember the drawn faces,
the dejected figures of my two men, and I remember my youth
and the feeling that will never come back any more – the feeling
that I could last for ever, outlast the sea, the earth, and all men;
the deceitful feeling that lures us on to joys, to perils, to love, to
vain effort – to death; the triumphant conviction of strength,
the heat of life in the handful of dust, the glow in the heart

that with every year grows dim, grows cold, grows small, and expires – and expires, too soon, too soon – before life itself.

And this is how I see the East. I have seen its secret places and have looked into its very soul; but now I see it always from a small boat, a high outline of mountains, blue and afar in the morning; like faint mist at noon; a jagged wall of purple at sunset. I have the feel of the oar in my hand, the vision of a scorching blue sea in my eyes. And I see a bay, a wide bay, smooth as glass and polished like ice, shimmering in the dark. A red light burns far off upon the gloom of the land, and the night is soft and warm. We drag at the oars with aching arms, and suddenly a puff of wind, a puff faint and tepid and laden with strange odours of blossoms, of aromatic wood, comes out of the still night – the first sight of the East on my face. That I can never forget. It was impalpable and enslaving, like a charm, like a whispered promise of mysterious delight.

We had been pulling this finishing spell for eleven hours. Two pulled, and he whose turn it was to rest sat at the tiller. We had made out the red light in that bay and steered for it, guessing it must mark some small coasting port. We passed two vessels, outlandish and high-sterned, sleeping at anchor, and, approaching the light, now very dim, ran the boat's nose against the end of a jutting wharf. We were blind with fatigue. My men dropped the oars and fell off the thwarts as if dead. I made fast to a pile. A current rippled softly. The scented obscurity of the shore was grouped into vast masses, a density of colossal clumps of vegetation, probably – mute and fantastic shapes. And at their foot the semicircle of beach gleamed faintly, like an illusion. There was not a light, not a stir, not a sound. The mysterious East faced me, perfumed like a flower, silent like death, dark like a grave.

And I sat weary beyond expression, exulting like a conqueror, sleepless and entranced as if before a profound, a fateful enigma.

A splashing of oars, a measured dip reverberating on the level of water, intensified by the silence of the shore into loud claps, made me jump up. A boat, a European boat, was coming in. I

invoked the name of the dead; I hailed: *Judea* ahoy. A thin
shout answered.

It was the captain. I had beaten the flagship by three hours,
and I was glad to hear the old man's voice again, tremulous and
tired. 'Is it you, Marlow?' 'Mind the end of that jetty, sir,' I
cried.

He approached cautiously, and brought up with the deep-sea
lead-line which we had saved – for the underwriters. I eased my
painter and fell alongside. He sat, a broken figure at the stern,
wet with dew, his hands clasped in his lap. His men were asleep
already. 'I had a terrible time of it,' he murmured. 'Mahon is
behind – not very far.' We conversed in whispers, in low
whispers, as if afraid to wake up the land. Guns, thunder,
earthquakes would not have awakened the men just then.

20

The Death of the Ship

Anyone who is given to wandering along estuary shores must have come across the bones of an old ship lying in the mud or in some rocky cove. He may have wondered what brought her there to die. Was it her old age, when she had reached the end of her time, that made men drag her to this last resting place? Was she still sound but outclassed by new designs, or overtaken by her engined rivals? If she is still clothed in her planking I will look more closely at her and wonder what she was, where she came from, and what was the nature of her business on the sea.

All ships must die, and Hilaire Belloc tells us about one he knew and loved.

A. R.

from '*ON SAILING THE SEAS*' by HILAIRE BELLOC

THE other day there was a ship that died. It was my own ship, and in a way I would it had not died. But die it had to, for it was mortal, having been made in this world: to be accurate, at Bembridge, in the Isle of Wight, nearly sixty years ago. Moreover, since boats also must die, it is right that they should die their own death in their own element; not violently, but after due preparation; for, in spite of modern cowardice, it is better to be prepared for death than unprepared.

They may tell me that a ship has no being at all; that a boat is not a person, but is only a congeries of planks and timbers and spars and things of that sort. But that is to open up the whole debate, undecided, branched out, inexhaustible, between realism and nominalism – on which I wish you joy.

She was my own boat, and I knew her very well, and I loved her with all my heart. I will offer you speculation on whether, now she has dissolved her being in this world of hers – which was sand and mud, salt water, wind and day and night and red and green lights, and harbours far away – she shall not be a complete boat again with all her youth upon her, in the paradise of boats. You may debate that at your leisure.

She had been patched up for years past. So are men in their old age and their decay. As the years proceeded she had been more and more patched up. So are men more and more patched up as the years proceed. Yet all those who loved her tried to keep her going to the very last. So it is with men.

But my boat was happier than men in this, that no one desired her death. She had nothing to leave, except an excellent strong memory of days calm, days windy, days peerless, days terrific, days humorous, days empty in long flats without a breath of wind, days beckoning, principally in the early mornings, leading on her admirable shape, empress of harbours and of the narrow seas. Also, she had no enemies, and no one feared her. There was no one to say, as there is of men, 'I shall be glad when they are out of the way.' There was no one to wish her that very evil wish which some men do other men – themselves evil: 'I am glad to think that he is dead.'

No. My boat went most honourably and straightly to her death. She had nothing to repent, nothing to regret, nothing to fear, nothing to be the cause of shame. It is so with things inanimate, and, indeed, with animals. It is so with everything upon this earth except man.

My boat was the best sea-boat that ever sailed upon the sea. The reason of this was that her lines were of the right sort, belonging, as they did, to the day when England was England; and my boat was so English that if you had seen her in any foreign port you would have known at once that you had seen an English thing. But, indeed, nowadays, what with their boats made like spoons and their boats made like table-knives, and their boats made like tops, and their boats made like scoopers, and their boats made like half-boats, cut away in the middle, no one can tell whether a boat is Choctaw, Esquimaux, or Papuan.

For boats have nowadays fallen into chaos, like everything else.

But this boat was plumb English, codfish nose and mackerel stern. She was, between perpendiculars, twenty-nine and a half foot. She was, over all, thirty-six foot. She was of cutter rig, she was nine tons – according to the only measurement worth having, and fancy measurements may go to the devil. Four men were happy on board her, five men she could carry, six men quarrelled. She did not sail very close to the wind, for she was of sound tradition and habit, the ninth of her family, and perhaps the last. To put her too close was to try her, and she did not like it. But she would carry on admirably four points off, and that is all you need in any boat, I think. She drew from just over to just under six foot, according to the amount of human evil there was aboard her and of provision therefor. And she never, never failed.

She never failed to rise to a sea, she never failed to take the stiffest or most sudden gust. She had no moods or tantrums. She was a solid, planted thing. There will be no more like her. The model is broken. There was a day when I should have cared very much. Now I am glad enough that she is gone down the dark way from which, they say, there is no return. For I should never have sailed her again.

He who had designed the lines of her approached the power of a creator, so perfect were they and so smooth and so exactly suited to the use of the sea. For modern men would have made her, no doubt, with a view to speed or with a view to holding this or that, or having this or that luxury aboard; to finding a place aft for the abomination of an engine. But those who made her knew nothing of all these things. They made her to be married to the sea.

As to speed, I suppose she never in her life made nine full knots in one hour. (As for those who say you cannot sail so many 'knots' in an hour, and that the expression is inaccurate, because a knot is not the same thing as a sea mile – my feeling about them is so strong that I dare not express it in words; so I leave it at that.) I say I doubt if she ever made nine knots in the hour, even on that famous day when she ran violently over-canvased because she had jammed a block, roaring from the flats east of Gris-nez to the flats of Romney in just over three hours, not knowing whither she went, nor I either until the land was suddenly upon us – as suddenly as the land had left us when we first rushed out into that thick weather – and that, God help me! was more than a quarter of a century ago.

In a good breeze, and behaving with comfort and tradition, she might make seven knots: or, again, one and a half. I have known her go out of Salcombe of a Saturday on the first of the ebb at noon, and make Torquay on the Monday morning with an oiled calm in between. On the other hand, she once ran me from that same Torquay to the Solent in less time than it takes a man to betray his loyalties or to deny his God: or, at least, in less time than it takes him to change his habits in the way of treason.

She once took me round from Dorsetshire to Cornwall one summer night and with a wind off the land which was much too strong in passing Bolt Head; and she has taken me here and she has taken me there; and now we are to part – if not for ever, at any rate for a good many weeks or months or years. Which things, I suppose, are inconsiderable to Eternity. No matter. We part.

The patching up had got more and more difficult. It had had to be renewed more and more often. The expense was nothing. We will always pay for doctors when it is a matter of those we love. But off the Norman coast the other day she gave me that look which they give us before they leave us, and she started a plank. It was high time. Had she not been near the piers it might have gone hard with those on board. But she got through, though the Channel was pouring in, and she reached the basin within, her cockpit half full, and then lay up upon the mud. And there she did what corresponds in man to dying. She ceased to be a boat for the purposes of a boat any longer. She was no-longer-patch-up-able. She had fulfilled her task. It was all over. She had taken to her repose.

Very soon she with hammer and wedge was dissolved into her original elements – all that was mortal of her – and the rest is on the seas of paradise.

I wish I were there – already: now: at once: with her.